FRED SANGER – DOUBLE NOBEL LAUREATE
A BIOGRAPHY

Considered 'the father of genomics', Fred Sanger (1918–2013) paved the way for the modern revolution in our understanding of biology. His pioneering methods for sequencing proteins, RNA and, eventually, DNA earned him two Nobel Prizes. He remains one of only four scientists (and the only British scientist) ever to have achieved that distinction.

In this first full biography of Fred Sanger to be published, Brownlee traces Sanger's life from his birth in rural Gloucestershire to his retirement in 1983 from the Medical Research Council's Laboratory of Molecular Biology in Cambridge. Along the way, he highlights the remarkable extent of Sanger's scientific achievements and provides a real portrait of the modest man behind them. Including an extensive transcript of a rare interview with Sanger by the author, this biography also considers the wider legacy of Sanger's work, including his impact on the Human Genome Project and beyond.

~

George G. Brownlee is Emeritus Professor of Chemical Pathology at the Sir William Dunn School of Pathology, University of Oxford, and a Fellow of Lincoln College, Oxford. He studied under Fred Sanger at the MRC Laboratory of Molecular Biology, Cambridge, where he later became an independent researcher. He is a recipient of the Biochemical Society's Colworth and Wellcome Trust Medals, a Fellow of the Royal Society and the Academy of Medical Sciences and is an EMBO Member. He received the Haemophilia Medal (France) and gave the Owren Lecture (Norway) for his pioneering work on haemophilia. He retired in 2008 to write this biography.

FRED SANGER DOUBLE NOBEL LAUREATE

A BIOGRAPHY

George G. Brownlee
Sir William Dunn School of Pathology
and Lincoln College, University of Oxford

CAMBRIDGE
UNIVERSITY PRESS

University Printing House, Cambridge CB2 8BS, United Kingdom

One Liberty Plaza, 20th Floor, New York, NY 10006, USA

477 Williamstown Road, Port Melbourne, VIC 3207, Australia

314-321, 3rd Floor, Plot 3, Splendor Forum, Jasola District Centre, New Delhi - 110025, India

79 Anson Road, #06-04/06, Singapore 079906

Cambridge University Press is part of the University of Cambridge.

It furthers the University's mission by disseminating knowledge in the pursuit of education, learning and research at the highest international levels of excellence.

www.cambridge.org
Information on this title: www.cambridge.org/9781108794534

First published 2014
First paperback edition 2020

A catalogue record for this publication is available from the British Library

Library of Congress Cataloging in Publication data
Brownlee, George G., 1942–
Fred Sanger, double Nobel laureate : a biography / George G. Brownlee, Sir William Dunn School of Pathology and Lincoln College, University of Oxford.
pages cm
Includes bibliographical references and index.
ISBN 978-1-107-08334-9 (Hardback)
1. Sanger, Frederick, 1918–2013. 2. Biochemists–Great Britain–Biography. I. Title.
QP511.8.S325B76 2015
572.092–dc23
[B]
2014021241

ISBN 978-1-107-08334-9 Hardback
ISBN 978-1-108-79453-4 Paperback

EDWARD GEORGE BROWNLEE (1973–1995)
He helped me prepare for my interview of Fred Sanger

Contents

List of figures

Foreword

In the 1950s, biology was seen as such a soft science that it was not taught in the boys' grammar school that I attended. Boys who wanted to study biology had to suffer the taunts of fellow pupils as they crossed the road for lessons at the high school for girls. That it is now a subject with a cutting edge (and fit for boys!) is due in large part to the scientific contributions of Fred Sanger. That such a modest, self-effacing man whose childhood nickname was 'Mouse' could have such a profound impact needed some explanation. In this biography, George Brownlee gives us that explanation.

When Fred Sanger joined the Laboratory of Molecular Biology (LMB) in Cambridge in 1962, he had already won a Nobel Prize for his work on the structure of insulin. This Nobel Prize was, appropriately, for chemistry, since he showed that proteins were proper chemicals, with a defined structure, and not some ill-defined mixture as thought by some. It was a discovery that opened up the chemistry of macromolecules, and created a new subject, distinct from the biochemistry of those days, that demanded a new name: molecular biology. In the LMB he turned his attention to the nucleic acids, no doubt influenced by the presence of Francis Crick's Division of Molecular Genetics. While the molecular geneticists treated their central subject as a coding problem – how does the information in DNA end up as a sequence of amino acids – and used clever genetic tools to crack the problem, Sanger carried on with his chemist's approach and developed powerful ways to sequence first RNA and then DNA. Along the way, as Brownlee describes, he made a number of important discoveries. And he introduced the notion that it was important to sequence entire genomes and then ask what the sequence told about the biology – an approach that is now called hypothesis-free discovery, carried out on a massive scale. But his work was not appreciated by all his colleagues. I once overheard Francis Crick remark: 'The trouble with Fred is that he has developed all these powerful methods, but we can't persuade him to do anything interesting with them.'

Crick's disdain echoed an attitude to technology development common in those days. It was seen as a poor second to what was done by those using technology to make discoveries or test ideas; perhaps because technology is ephemeral while discoveries endure.

But Sanger didn't just want to apply his methods to answer questions, he wanted to develop new methods to tackle ever more difficult molecules – that was his thing; as popular parlance would have it, it was in his DNA. Wisely, Brownlee avoids the question of whether it was really in his DNA, whether his intelligence and inventiveness were genetically programmed, and seeks the explanation in his character, as shaped by his upbringing in a Quaker family. What qualities are needed to be an outstanding inventor? What we see in Brownlee's description of Sanger are imagination and exceptional boldness, determination to the point of doggedness, and strong self-belief. Once he had decided on a problem, Sanger went to extraordinary lengths in his drive to achieve his objectives. I learned one of his RNA sequencing techniques from Ken Murray, who had worked with Sanger. Among many other steps, the method involved electrophoresis on large sheets of DEAE paper, at 2000 V, in tanks holding gallons of highly flammable white spirit – to the list of qualities, we should add almost reckless courage!

Whatever the source of his qualities, he certainly needed them when, in 1970, he came to tackle the problem of sequencing DNA: 'the ultimate challenge'. It is a remarkable story, fully described in Brownlee's account. We see it unfold over a period of six or seven years with Sanger slowly edging forward in significant steps: building on his experience of separation methods that he developed in his work with RNA; adopting advances made by others, such as the application of chain extension by polymerase; and, ultimately, finding the solution by combining chain terminating precursors, which he had to synthesise himself, with gel electrophoresis.

Sanger himself doesn't attribute his successes to any personal quality. When Brownlee asks, 'What do you think are the most important things, in general, that allowed you to make these huge advances in molecular biology…?' Sanger replies, '… being at the right place at the right time … I certainly have been lucky in being able to be independent … Maybe I've had a few ideas but I never know where they come from … they may come from talking to other people around the lab.' This may be one occasion on which we catch the great man being less than fully truthful.

The enormous impact of Sanger's sequencing method is so wide in its scope that it is impossible to capture in a short biography. Brownlee illustrates it with three telling examples. First, the sequencing of the human genome. The success of this enormous project stimulated the second example, the foundation of a DNA sequencing industry with the introduction of automation and high-throughput methods. There are now numerous companies manufacturing and selling machines and reagents and others that offer sequencing services. And finally, the most important impact that DNA sequencing has had on medicine, the discovery of genes and mutations associated with cancer.

Sanger, typical of many successful scientists, is reserved, even shy. So, we are fortunate that he offered to be interviewed by Brownlee and to provide an account which ranges wide over his early life and his scientific career. Brownlee includes a verbatim transcript of this interview. This is a special treat for those of us who knew both men: we can picture this conversation between two colleagues and friends, see their faces and hear their voices. More important, perhaps, is the record it embodies. It will be a priceless source for those researching the broader history of the remarkable period that saw the birth of modern biology. The biography will provide inspiration and encouragement for aspiring inventors, for whom there is still a great need. Above all, it provides lessons for the funders: to find the next Sanger, they must be prepared to give long-term support with strings loosely attached.

Sir Edwin Southern
Professor of Biochemistry (Emeritus) and Fellow, Trinity College,
University of Oxford

Acknowledgements and original sources

The following are major sources in addition to Fred Sanger's own publications (see Bibliography):

1. Biochemical Society video archive (1987). Sanger-sequences [Dr F Sanger interviewed by Mr H Judson, 13 Nov 1987].
2. Biochemical Society video archive (1992). A life of research in the sequences of proteins and nucleic acids: Dr F Sanger OM, CH, FRS interview with Professor George Brownlee FRS, 9 Oct 1992.
3. Finch J (2008). *A Nobel Fellow on Every Floor*. Cambridge: Icon Books.
4. Silverstein A, Silverstein E (1969). *Frederick Sanger: The Man Who Mapped Out a Chemical of Life*. New York: John Day.
5. Wellcome Archive, Euston Road, London. Fred Sanger's laboratory notebooks form part of the Biochemical Society collection held at the Wellcome Archive (SA/BIO/P/1).
6. Wellcome Beit Archive, Euston Road, London. Fred Sanger's Beit Fellowship archive (SA/BMF/A/2/231).

Original sources

Fred Sanger encouraged me to write his biography when I first asked him in 2008. A year later he told me to 'get a move on', if it were to be published before he died (I did not succeed in this). His wife, Joan (deceased 2012) told Sue, my wife, in 2010 to 'make Fred's biography interesting'. Fred himself, who died in 2013, was not well enough to be able to check the manuscript for me, although it was written before he died. His son, Peter Sanger, read the section dealing with the family history and has given me significant help and encouragement by lending me original images from the family archive, and some letters and drafts of speeches that Fred Sanger gave after his retirement in 1983. Peter Sanger also allowed me to interview him about Fred as a father and a family man.

I have also had generous help from Mary (May) Willford, *née* Sanger, Fred Sanger's younger sister, and her sons Noel and Julian Willford. May has given me information about Fred as a person, helped me understand the Sanger family tree and lent me Cicely Sanger's (Fred's mother's) diary [or children's book] from 1918 to 1937. I had access to a visitors' book (held by May Willford) originally kept by the Sangers' parents (Frederick and Cicely Sanger) covering the period 1918 to 1938 with further entries from 1938 to 2001. May Willford and her sons Noel and Julian also brought the Svenhonger Diary, second volume, 1941, 1942, 1943 (a private publication by Clyde Sanger, about 2000) to my attention. These are the diaries of Gerald Sanger, Fred Sanger's cousin, son of Fred's Uncle William. May Willford also lent me a DVD of an interview Fred Sanger gave at his old school, Bryanston, in 2007, when he opened the new school Sanger Building.

Fred Sanger's laboratory notebooks from 1940 to 1983 and limited additional material are archived for the Biochemical Society at the Wellcome Library, Euston Road, London and I thank the archivists for access to this archive. I also accessed part of the Beit/Sanger archive, also held at the Wellcome Library (courtesy of the archivist).

I thank numerous scientists whom I interviewed in connection with this biography, namely Jerry Adams, (the late) Richard Ambler, Elizabeth Blackburn, Alan Coulson, Ted Friedmann, Brian Hartley, Celia Milstein, (the late) Sir Kenneth Murray, George Petersen, Julia Porter and John Sedat. I corresponded with John Donelson to clarify his role in the development of the gel 'read-off' methods in the period 1971–3 when he was a postdoctoral fellow in Fred's laboratory. All their individual recollections have given me many insights into Fred Sanger's character and personality. I thank my wife, Sue Brownlee, for help at my interviews of scientists, since she knew many of them in her own right, but also for all the help and patience she has shown while I was writing this biography. Her memory of events often complemented mine and has added significantly to this biography.

I interviewed Hagen Bayley, University of Oxford and David R. Bentley, Illumina, Cambridge, in connection with Chapter 7 on post-Sanger sequencing. I particularly thank David Bentley and Larry McReynolds, who critically reviewed Chapter 7. Sir Michael Stratton, Dr Lisa Walker and Sir Ed Southern suggested corrections and improvements to Chapter 8. I am most grateful to Ed Southern for writing a

Foreword to this biography. I thank Paul Berg, Elizabeth Blackburn, Sir John Sulston, David Bentley and Sir Paul Nurse for providing illuminating commentaries. I am grateful to the author Colin Tudge for his encouragement while I was writing this biography.

The archivist, Annette Faux, at the Medical Research Council Laboratory of Molecular Biology, Cambridge, has generously provided me with original images from their archives (copyright 'MRC Laboratory of Molecular Biology'). Dr John Lagnardo, as the Biochemical Society archivist, has given me significant help and permission to reproduce images published in their Biochemical Society video archive of Sanger in 1992 (The Biochemical Society, with permission). In many but not all cases, I have had access to the original image used for this video, courtesy of Peter Sanger, and have used original images in this biography where possible, because of their higher resolution. I thank the librarian, Joanna McManus, *née* Hopkins, at the Royal Society London, and Michael Gurr, Hans Tuppy, (the late) Ted Thompson and Stephen Burch who have helped me with individual images. Every effort has been made to track down original copyright owners. I would be grateful for information on any omissions, so that corrections can be included in any future biography or reprinting.

Only limited correspondence between Fred Sanger and his scientific colleagues survives. What is available is thanks to George Petersen and the late Sir Kenneth Murray who made their correspondence with Sanger available to me. I have not been able to trace Sanger's known extensive scientific correspondence between 1962 and 1983 during his period at the Laboratory of Molecular Biology, or any of his correspondence while in the Biochemistry Laboratory at the University of Cambridge between 1940 and 1962. I would be grateful for any information that might help locate this missing correspondence.

Brief chronology

1918	Born in Rendcomb, Gloucestershire, 13 August, second son of Frederick Sanger, MD and Cicely Sanger, *née* Crewdson.
1927	Entered Downs Preparatory School, Malvern, as a boarder. This was a Quaker school.
1932	Entered Bryanston School, Dorset – a liberal, public (fee-paying) school as a boarder.
1936	Went up to St John's College, Cambridge, to read Natural Sciences. He took four years overall, because he needed an extra year for Part I. He specialised in biochemistry in his fourth year. Father, Frederick, died in 1937; mother, Cicely, died in 1938.
1940	Started PhD in Department of Biochemistry, Cambridge, supervised initially by Bill Pirie, then by Albert Neuberger. Self-funded. Married Joan Howe in December 1940.
1943	Submitted PhD thesis, University of Cambridge: 'The metabolism of the amino acid lysine in the animal body'.
1944–55	Sequence of insulin, supported by a Beit Memorial Fellowship from 1944 to 1951, and by the Medical Research Council from 1951 onwards.
1958	First Nobel Prize in Chemistry.
1962	Moved to Medical Research Council's Laboratory of Molecular Biology, Cambridge.
1962–70	RNA sequencing.
1970–80	DNA sequencing.
1980	Second Nobel Prize in Chemistry, jointly with Paul Berg and Walter Gilbert.
1983	Retires aged 65.
1983–2013	Retirement: gardening, boating, reading, seeing grandchildren.
2012	Joan Sanger, Fred's wife, dies.
2013	Fred Sanger dies.

Honours

Civic

1963 Commander of the Order of the British Empire (CBE)
1981 Companion of Honour (CH)
1986 Order of Merit (OM)

Scientific

1951 Corday–Morgan Medal and Prize, Chemical Society
1954 Fellow of King's College, Cambridge; 1983, Honorary Fellow
1954 Fellow of Royal Society
1958 Foreign Honorary Member of American Academy of Arts and Sciences
1958 Nobel Prize in Chemistry
1961 Honorary Member of American Society of Biological Chemists
1961 Member of the Academy of Science of Argentina
1961 Member of the Academy of Science of Brazil
1961 Honorary Member of the Japanese Biochemical Society
1961 Corresponding Member of the Asociación Química of Argentina
1962 Member of the World Academy of Art and Science
1966 Alfred Benson Prize, Denmark
1966 Honorary Fellow National Institute of Sciences of India
1967 Foreign Associate of US National Academy of Sciences
1968 Honorary DSc, Leicester University
1969 Royal Medal, Royal Society
1970 Honorary DSc, University of Oxford
1970 Honorary DSc, Strasbourg University
1971 Sir Frederick Gowland Hopkins Memorial Medal, Biochemical Society

1971	Gairdner Foundation Annual Award, Canada
1976	William Bate Hardy Prize, Cambridge Philosophical Society
1976	Hanbury Memorial Medal, Pharmaceutical Society of Great Britain
1976	Fellow of the Royal Society of Edinburgh
1977	Copley Medal, Royal Society
1978	G. W. Wheland Medal, Chicago University
1979	Louise Gross Horwitz Prize, Columbia University
1979	Albert Laskey Award, New York
1979	Gairdner Foundation Annual Award, Canada
1980	Biochemical Analysis Prize, German Society of Clinical Chemists
1980	Nobel Prize in Chemistry
1981	Foreign Associate, French Academy of Sciences
1982	Corresponding Member, Australian Academy of Sciences
1983	Gold Medal, Royal Society of Medicine
1983	Honorary DSc, University of Cambridge
1994	Association of Biomolecular Resource Facilities Award
2010	Honorary Fellow, St John's College, Cambridge
2013	Fellow, American Association for Cancer Research Academy

Introduction

Fred Sanger asked me to interview him for the Biochemical Society archives in October 1992. He never told me why he chose me as an interviewer rather than one of his many colleagues and friends in Cambridge. I had worked with Fred in Cambridge from 1963 to 1980 but by 1992 had moved to the Sir William Dunn School of Pathology at the University of Oxford. Naturally I was honoured and accepted. In that interview Fred Sanger gave a full and frank account of his life and provided some insight into what qualities he thought were needed to be a successful scientist. This biography is, in part, based on that interview.

Fred had an unusual upbringing and it may surprise many readers to learn that he was not initially committed to research. As a young man he had a strict Quaker upbringing. He was ambivalent about studying medicine, changing his mind to study biochemistry rather than medical subjects at Cambridge because his father, a doctor in general practice, was always rushing around attending to patients and 'could not really concentrate on anything'. Both his parents died when he was an undergraduate, leaving him vulnerable. He doubted he was good enough to do research and applied at the last minute to study for a higher degree, a PhD, in biochemistry at Cambridge only after he learnt he had been awarded, to his surprise, a first class degree. Sanger was a conscientious objector in the Second World War. He learned to do research rather than killing the enemy.

In spite of this uncertain beginning to his career, he quickly showed aptitude for research. He seemed to have the ability to succeed in solving 'impossible research problems' where others had feared to tread. He was a 'hands-on' researcher doing research himself with the help of a trusted technical assistant right up to the end of his career. Many of his critical research findings were the result of his own findings carried out personally. This did not mean he was not a good leader. In fact he was a consummate team leader and early on attracted others to his research team because of his achievements and the great personal effort he made. This personal commitment to research inspired confidence in his many collaborators, including me. He was also unusual in stressing the contribution of the whole of his research team. In particular, he emphasised the role of his two technicians, Bart Barrell and Alan Coulson, as much as the effort of his more academically qualified collaborators. Fred Sanger did not allow himself to be distracted by teaching or scientific administration. He was 100% committed to research.

A modest man, Fred Sanger's name deserves to be better known for inventing methods for sequencing proteins, for sequencing RNA and especially for the Sanger method of sequencing DNA – the genetic basis of life. More or less by chance he made many other highly significant and surprising discoveries in molecular biology, for which he was justly proud. He was the first to directly confirm the genetic code, the first to discover the unexpected phenomenon of 'overlapping genes' and the first to show that the genetic code could vary in different organisms. The Sanger 'dideoxy' method was used to unravel the human genome sequence in 2000. It is difficult to think of a contribution to modern biology that is more significant.

I was fortunate to work in Fred Sanger's laboratory in Cambridge from 1963 to 1980, first as his PhD student and later as an independent researcher. At the interview for my PhD Fred asked me if 'I wanted to sequence proteins or RNA'. I replied that I wanted to sequence RNA. My decision was wise as progress was rapid in the subsequent decade in developing new methods for sequencing RNA. This RNA sequencing phase – perhaps the least well-known period of Fred's research – was a transition period after Fred had worked out how to sequence proteins but before he had developed rapid methods for sequencing DNA. Fred had chosen RNA rather than DNA because the only known small nucleic acids available were RNA. DNA was too long and completely out of range at that time in the early 1960s.

This biography traces Fred's life from his birth in 1918 in a remote village in Gloucestershire to his retirement from the justly famous Laboratory of Molecular Biology in Cambridge in 1983. Fred spent his whole scientific career in Cambridge and must be one of Cambridge's best-known sons. I describe Fred Sanger's upbringing and scientific achievements in Chapters 1–3 followed in Chapters 4–6 by a slightly edited transcript of my interview with him in 1992. Unusually for a biography I include two further chapters to illustrate the ongoing impact of Fred's DNA sequencing method that he described in 1977. Chapter 7 – post-Sanger automated sequencing – describes the development of his method that others used to unravel the human genome sequence of nearly 3×10^9 bases (or 3 000 000 000 bases) in the year 2000. Chapter 7 brings sequencing up to date, describing current newer sequencing methods using massively parallel DNA sequencing. Sanger always wanted to make an impact in medical research. Chapter 8 describes results of one current medical-research-based sequencing project to identify the genetic basis of breast cancer. In my opinion these rapid-sequencing methods of cancers – still in part based on Sanger's method – describing the genes that are actually mutated in cancers will be the main driver for new drugs specifically targeted to individual genes in cancers. Finally, Chapter 9 includes commentaries by five eminent molecular biologists describing, each in their own way, Fred Sanger's legacy to science.

My hope is that his main scientific achievements are adequately described in a way that is understandable by the interested general reader. I include endnotes and a full list of Sanger's scientific papers for readers who want more details, but my overriding aim is to try to answer the question as to why Fred was able to contribute so much to molecular biology in one lifetime. How was he able to work out how to sequence a protein, insulin, from 1944 to 1955 when the problem of protein sequencing was regarded as impossible by other scientists at the time? How did he have the foresight to take on the challenge of sequencing RNA in 1962 and then to succeed in developing a rapid method for DNA sequencing by 1977? Was it his upbringing at home and the influence of his medical father? Was it his training at the well-known public school Bryanston, or as an undergraduate in Cambridge? Was it the influence of the Cambridge biochemistry department where he did his PhD under Albert Neuberger, and where he would have met other visionary scientists? Was it his interpersonal skills perhaps based on his upbringing as a Quaker? Was it perhaps in his DNA?

Fred is amongst only a handful of people ever to have ever been awarded two Nobel Prizes. Such prizes are awarded by the Swedish Academy of Sciences annually to people who have made outstanding contributions to knowledge, or to peace in the case of the Nobel Peace Prize, and are widely considered to be the ultimate accolade. To gain one Nobel Prize might be considered by critics to be lucky or because the person was in the right place at the right time. To gain two such prizes cannot be attributed to chance.

1

A Quaker upbringing

Biography is perhaps one of the most difficult of all historical and literary ventures, because it involves not only the compilation of material and its understanding, but also an attempt, which can never be wholly satisfactory, to enter into the soul of the subject and to create an honest account of life as well as to present a portrait.[1]

Cicely Crewdson, an elegant lady in her late thirties, was on holiday with her father at their family holiday home in the small village of Syde in the Cotswolds. Her holiday was nearly spoiled because she needed urgent treatment for a septic finger. Sepsis was potentially life-threatening in those days before effective antibiotics, such as penicillin, were available. The local doctor was summoned and agreed to come over from the nearby village of Rendcomb to treat Cicely. The doctor's name was Frederick Sanger and he was still a bachelor. Her treatment would have needed a number of visits by this doctor to see how the patient's finger was improving. Cicely Crewdson and Frederick Sanger got to know one another and doctor and patient were married in 1916.

We can only guess what attracted Frederick and Cicely to one another but Frederick would have been a quite suitable match for Cicely. Educated at St John's College, Cambridge, he qualified as a doctor and completed his MD in 1902. Soon after he travelled to China as a missionary where he worked as a hospital doctor and found time, energy and enthusiasm to set up a new school for poorer children who were generally denied the education available for the children of the mandarin or upper-class families. Returning to Devon, his parent's family home – probably for health reasons in 1912, Frederick then moved with his widowed mother, Ann, to Gloucestershire where he practised as the local doctor. His mother died shortly after the move in 1913.

Cicely Crewdson was the youngest daughter of a rich and successful family, who owned a cotton mill in Styal, Cheshire near Manchester. The Crewdsons had strong Quaker traditions, although Cicely's father, Theodore, had actually renounced Quakerism in favour of the Church of England. Cicely, the youngest of six children, had been brought up as a Victorian lady of the Church of England persuasion. There was no question of her going out to work. Frederick Sanger was nearly 40 years old and Cicely Crewdson 36 years old when they married in 1916, so they were ready to settle down and have a family.

Cicely and Frederick set up home in The Old House, Rendcomb in Gloucestershire with its large garden backing onto the River Churn. Rendcomb is a small, picturesque Cotswold village that today is probably largely unchanged since 1916. The village has a parish church with a twelfth-century font, a village store and a manor house – converted to a school, along with a small number of typical Cotswold stone-built houses with attractive stone-tiled roofs, in the pleasantly undulating countryside of the Churn valley. Sheep would be grazing on the hillside then as they are today in this predominantly farming community. It would have been an isolated country living then, quite remote from the larger cities. Few local people then would have been able to afford cars. Today communications are better but the local roads are still narrow and winding, discouraging any through traffic in the village. The nearest market town is Cirencester, built in Roman times, 6 miles away.

Frederick had his one-room, quite claustrophobic surgery 'outhouse' at the side of the house, where patients would be seen. In those days doctors did their 'rounds' and often visited their patients in their own homes so there was no need for a large surgery. The small surgery contrasted with the large old house with its imposing frontage, Cotswold stone tiles and large reception area inside. Cicely had to run the house, but she would have been no stranger to this as she was brought up in a substantial house in Styal, Cheshire. No doubt, her older, unmarried sister May, who often visited Rendcomb, sometimes with her father Theodore

Crewdson, would have given advice on securing the services of servants. Cicely, like her parents, would have had the help of a housekeeper, a cook, a gardener and later a nanny when the children were born.

Cicely and Frederick Sanger had three children. Theodore (Theo) was born in 1917, Frederick (Fred) – the subject of this biography – was born soon after in 1918 and Mary (May) five years later in 1923. The three children, Theo, Fred and May, were brought up as Quakers as father Frederick had converted from the Church of England after his marriage, influenced by the books and traditions in the Crewdson family. Mother Cicely remained Church of England and attended the local Anglican church every Sunday. Father Frederick went to a Quaker Meeting House with the children. The children were taught the importance of telling the truth, and prayers were said every morning and night. Fred, the second child, was strongly influenced by his elder brother, Theo, and they played together a lot. Theo was more extrovert than Fred, who described himself later as a 'quieter and more retiring fellow' than his elder brother Theo.

Cicely and Frederick were well matched and happily married. They enjoyed taking walks together in the local countryside and going riding. Frederick was a romantic and wrote love poems to his wife, Cicely, in a neat script, long after they were married. Cicely was kindly and more serious than Frederick, who had more of a sense of humour. There were never any arguments in this family even if there were differences of opinion.[2] It was a happy, carefree time. There were no money worries and it was a wonderful, privileged upbringing for Fred and his elder brother and younger sister.

Cicely kept a diary from 1918 to 1937 detailing the major events in the family life over these years.[3] Shortly after Fred was born on 13 August 1918, Cicely notes in her diary a difference in temperament between Fred and his elder brother, Theo:

> He is a dear little fellow, but very small 5¾ lbs at birth, and lost ½ lb the first week, but is gaining now. He is very quiet and peaceful, sleeps all day and will hardly wake up for meals, he opens his eyes at night. He is not a bit like Theodore . . .

Cicely's diary gives us many insights into Fred's temperament and upbringing as we shall see later. It is hard to ignore this insight – one wonders if she would have guessed that her second child would become a famous scientist. Sadly Cicely was never to know this as she died from cancer in 1938 when Fred was still an undergraduate at Cambridge. At Theodore's first birthday party in 1918 a cake was provided 'and it was iced – an unheard of thing in war time', Cicely records. Apart from this reference to icing on Theo's birthday cake, there is no mention of the First World War in Cicely's diary. With an estimated 2 million lives of servicemen and women lost in the UK in the First World War it would have been surprising if the wider Sanger family had not lost a relative. In fact Fred Sanger lost a distant second cousin, Henry. Perhaps their Quaker beliefs, that one should not kill others, kept the Sanger family partly insulated from the terrible war with its huge loss of life going on around them in Europe.

The Sanger family moved house in 1923 to 'Far Leys' in Tanworth-in-Arden, another village closer to Birmingham, when Fred was 5 years old. The reason was partly so that Frederick, his father, could be nearer Birmingham, an important Quaker centre. Fred's early schooling was taken care of by a Quaker governess, Miss Shewell, who taught Fred, Theo and two other village children until Fred was 9 years old.

The governess, Helen Mary Shewell, or HMS (Her Majesty's Ship!) as she was called by the children, replaced the first governess, Miss Potter, on 18 May 1924. Miss Potter, according to Cicely's diary, 'certainly got the boys on well in lessons, but I did not like her way with them. They got frightened of her and disliked lessons so I felt we must make a change.' With Miss Shewell they are now 'much enjoying lessons. She stays three afternoons a week and takes them out and they have lovely times in the sand heap with her. She thinks them quite forward for their age. Der (Cicely's nickname for Fred) is very good at sums – both addition and subtraction.'

Miss Shewell taught both Theo and Fred at the same time and noted that Fred was brighter than his elder brother: 'Der (Fred) is the cleverest, he goes quietly on without any effort. Ogo (Theo) rushes at things and then often gets tired.'

Miss Shewell travelled each day from Hall Green, Birmingham to teach all three Sanger children and various village children, who came to the school room at 'Far Leys', for 10 years. She only left the Sanger home in 1934 when May was sent to Overstone – a boarding school for girls. HMS had become a family friend and Cicely wrote in her diary in September 1934:

> It is terribly quiet at home with no child. It seems so strange with no school room and Miss Shewell after ten years. I am glad she has got a post and hope we shall still see a lot of her. She certainly has done very well for my children.

The Sangers' visitors' book confirms that Miss Shewell, still living at Hall Green, Birmingham, remained a family friend since she stayed at the family's Caudle Green house in August 1939 for two nights, after both Frederick and Cicely had died.

Theo got Fred interested in science at that stage, as Theo was very keen on natural history – animals, looking at bugs, catching newts, etc. 'Far Leys' had a spacious garden and pond, which was full of fish, and in the spring masses of toads came in and spawned there. They also collected grass snakes, which although frightening to young children were harmless.

The next stage in the education of the young Fred, only 9 years old, was quite traumatic. He was sent away to the Downs School, Malvern (a Quaker school), as a boarder in 1927. Not liking this much at first, he slowly learned to adapt to the move away from his mother's apron strings and survive the inevitable bullying. Fred found schoolwork easy there. Cicely writes in her diary: 'Der (Fred) finds work in the 1st form easy, he is top of the form and I am afraid he will slack, but hope not.' His main interests at that time seemed to be centred around his school holiday activities. At about that time he took up carpentry and painting both encouraged by his parents. Cicely notes in her diary in February 1928 when Fred was 10 years old, 'The boys were both much occupied in the holidays in making secrets for my birthday. Der has had a most original idea and carved me a church on a small piece of wood and then coloured it. It is most effective and quite his own idea.' Later on the young Fred took up welding. He had a forge at home and made a garden gate for his parents. Fred was still very quiet at home – he was nicknamed 'Mouse'. His mother worried that he was not outgoing enough. In 1930 she wrote: 'Fred has come on lately. He looks much stronger, has grown and is fatter and has been very fit and well all this year. He is also not quite so shy. He is a most delightful quiet companion and is always ready to help and do things.'

In 1932 – now 14, Fred then moved as a boarder to Bryanston in Dorset, a public (private) school that was quite new then. Fred tried for a scholarship but failed to get one, records Cicely. 'Freddy finished up well at the Downs. He got his cricket colours but did not manage a scholarship to Bryanston; he and a friend tried for it but neither got it.' Bryanston suited Fred. He was happy there – perhaps by now he had adapted to being a boarder away from home, even though 'the food was pretty dreadful, certainly not as good as at the prep school I had been to, but there was enough of it'.[4] Fred thrived on the liberal, project-based teaching that had been introduced there. He also started learning some science from Geoffrey Ordish, the chemistry master, and Frazer Hoyland, the biology teacher. These masters were stimulating teachers who involved Fred, and presumably other pupils, in other after-school activities (e.g. biology club, chemistry club). Fred was a diligent pupil, progressing well, near but never at the top of the class. Fred participated in sport at Bryanston (as did all pupils) in rugby, cricket, swimming, fives and squash, but did not excel in these activities although he enjoyed them. He was more academic than sporty. Fred did very well in his School Certificate exam, and was labelled 'seven-credit Fred'. It is quite clear that Fred was hardworking and interested in his class work and did particularly well at chemistry and biology. This qualification guaranteed his entrance to St John's College, Cambridge where both his father and uncle Tom (Hubert) had been undergraduates.

Cicely and Frederick, father, had been hoping Fred would study medical subjects in the Natural Sciences Tripos when he went up to Cambridge, but Fred had decided otherwise, records Cicely in 1935:

> We are all delighted that Fred has got seven credits in the School Certificate, including Latin, so he is right for Cambridge now. He has decided he does not want to be a medic which is rather disappointing to us but we feel it is no use to press him if he doesn't like the idea.

Fred's last year at Bryanston was very pleasant and academically undemanding, since Fred had already secured his place at St John's. Fred did not try for a scholarship for Cambridge since he was advised by the physics master at Bryanston that 'he was not quite up to it'. In that final year in school in 1936 Fred went on a school exchange visit to Salem in southern Germany as the two schools had arranged exchange visits. Fred spent a couple of months there with fellow pupil David Forbes, enjoying the outdoor activities and learning some German. He was also exposed to the Hitler Youth Movement and refused to salute

'Heil Hitler' in class, but appears not to have been aware, like many others, of the impending threat of war.

On arriving in Cambridge in 1936 Fred had to decide which subjects to take for Part I of the Natural Sciences Tripos. It was clear he should take chemistry but he chose physics as another full subject, and maths and biochemistry as half subjects. His choice of biochemistry was prescient and arose from his meeting with Ernest Baldwin, who was a tutor at St John's College. He persuaded him to take biochemistry – a subject new to Fred that tried to explain 'biology in terms of chemistry'. Baldwin was an enthusiastic teacher and a member of the Biochemistry Department at the University of Cambridge. Gowland Hopkins had made this department pre-eminent in this subject because of his discovery of vitamins.

Fred did not care for his first St John's 'digs' (lodgings) in Bridge Street, Cambridge and he moved nearby to Park Parade off Jesus Green later, and was happier there. 'Fred is having rooms in Park Parade as he feels Bridge Street is so shut in. I hope he will keep fitter there as he always gets a cold as soon as he gets to Cambridge,' writes Cicely. She continues: 'Both boys [Theo was a new undergraduate at Trinity Hall at the same time as Fred was at St John's] find they have to work very hard to keep up. The Friends (Quakers) have been very good in looking after them and they both go to meetings.' The Quaker family tradition was seemingly a help to both Fred and Theo in settling down in Cambridge.

Fred's decision to forgo medical subjects in favour of basic science subjects caused him difficulty in Part I of the Tripos – a situation he had not expected. It turned out he needed three years to gain a second in his Part I Tripos, instead of the normal two years. The reason was that he found physics and maths very difficult, not having studied them beyond the General Certificate stage at school. He could not catch up with the other undergraduates, who had studied physics and maths for Higher Certificate at school. There was nothing for it. Fred admitted defeat with physics and gave it up after the first year. Instead he changed to physiology as a full subject in his second year, finding it much easier and more interesting than physics, probably because of his strong biological background at school. He struggled on with maths as a half subject but did not enjoy this subject. He liked biochemistry most and this was the subject he chose for Part II of the Tripos in his final fourth year as an undergraduate.

There is more for a young man to do in Cambridge than simply work. Fred was still a committed Quaker. He had joined the Cambridge scientists' anti-war movement and had signed the 'Peace Pledge Union', pledging that he would not fight. As a Quaker he was a confirmed pacifist. The Second World War started in September 1939 with the invasion of Poland by Germany, while Fred was still an undergraduate, about to start his fourth year at Cambridge specialising in biochemistry. His pacificism had two important consequences. He met his future wife, Joan Howe, a pretty brunette at Newnham College, hailing from near Leicester, through the anti-war group. They were married a year later in 1940 in Syde in the Cotswolds where his uncle Dilworth had a farm near to Rendcomb where Fred had lived as a child.[5] Sadly, both Fred's father and mother had died from cancer by this time. Fred also registered as a conscientious objector, which meant he was not called up to fight in the Second World War.

Fred had made no plans to continue an academic career until he heard he had been awarded a first for his biochemistry Part II. He decided, only then, that he was good enough to do research. Fortunately he did not need a sponsor for his PhD studies since he had become financially independent through an inheritance from his parents. So he was able to finance his own PhD studies at Cambridge. Even then there were difficulties to be overcome, because Bill Pirie, with whom Fred had initially chosen to work, decided to leave the department only one month after Fred had started his PhD, leaving Fred in the lurch. Fortunately Albert Neuberger then agreed to supervise Fred's research on the metabolism of lysine, an important constituent amino acid of proteins.[6]

2

How about studying insulin?

Fred Sanger started his research in arguably the best-known biochemistry department in Britain in 1940 under the leadership of Frederick Gowland Hopkins. Hopkins, affectionately known as 'Hoppy', held the first Chair of Biochemistry in Cambridge from 1914 and had modernised biochemistry in the UK. Biochemistry had lagged behind advances in this subject in Europe, and particularly in Germany, in the previous century. The subject had evolved from physiological chemistry and was originally a branch of physiology with a chemical bias. But Hopkins, as the first Professor of Biochemistry in Cambridge – although not the first in the UK which was in Liverpool University in 1902[1] – established a vibrant British school of biochemistry even though he had an unconventional training as an analytical chemist and medic and had never formally studied biochemistry himself. 'In Hopkins two things were significantly combined: the training and tastes of an organic chemist, and the imagination of a biologist and physician.'[2] A careful and committed experimentalist, he had brought Cambridge fame through his discovery of vitamins as accessory food substances needed in minute quantities, for which he was awarded the Nobel Prize in 1929. But his engaging, friendly and benign personality meant that 'Hoppy' headed an open and enquiring department, where no subject, including politics, was off-limits. He was popular and approachable with the advanced Part II students. He had appointed talented and sometimes controversial scientists, such as J. B. S. Haldane, Joseph Needham,

Malcolm Dixon, Bill Pirie, Ernest Baldwin and Robin Hill. Among his appointments were the women scientists Dorothy Moyle (later Needham) and Marjorie Stephenson who became eminent in their fields. Hans Krebs and Ernst Chain – fleeing persecution from Nazi Germany and both to become future Nobel prize winners – were made welcome by Hopkins in the 1930s. In fact the department was affectionately known by some as the 'Hopkins Matrimonial Agency' because of the number of marriages between research workers[3] and 'Little Moscow' because of the left-wing tendency of the Needhams, Pirie and others.[4] But science flourished in that environment, although by 1940, when Fred Sanger started his PhD, Hopkins was 'very old, rather frail and deaf, but he used to totter around the lab'. Nevertheless he talked to the young Fred Sanger and congratulated him on his first class degree.

Fred's research for his PhD was very much on the theme of essential factors in the diet – a major theme of Hopkins' own research. Fred, under the watchful eye of Albert Neuberger, who was relatively young then and still very active experimentally in the lab, was asked to investigate the breakdown products of lysine, when this amino acid was fed to young rats. Lysine was important because it was an 'essential amino acid' in the diet that was absent in some proteins. Like tryptophan, valine and some other amino acids, it could not be synthesised in the body from simpler chemicals *de novo*. Fred never achieved his objective of understanding the breakdown of lysine and said of his own thesis: 'I wouldn't say it was a great thesis.' Lysine metabolism was only understood many years later in the 1970s,[5] but Fred did manage to publish five papers from this period of research with Neuberger,[6] including some war work on the protein composition of various varieties of potatoes, although this was not of any real practical benefit to the war effort.

Fred describes Neuberger as his main teacher, the person who really taught him how to do research. 'He was a very kind person too and very helpful to me. I feel I owe a lot to him.' This may be 'unduly generous' according to Neuberger[7] although many scientists later in their careers recognise the importance of their first research mentor. Perhaps the real benefit of Sanger's PhD training under Neuberger was to introduce him not simply to the practical aspects of science – after all Fred was very practical and ingenious in making things himself. Rather Neuberger taught him the importance of tackling novel and important scientific questions and testing them in a methodical way. Although Fred was not successful in solving the ambitious aims he had been set by Neuberger, Fred had been tackling an important problem, rather than simply dotting the i's and crossing the t's on reasonably well-known work. Fred's later research on insulin, RNA and DNA demonstrated he always aimed high in his research and Neuberger must take some of the credit for this.

Fred's PhD thesis demonstrates his own critical thinking early in his career. For example, he was not able to experimentally reproduce, and by implication criticised, some early work of Hans Krebs with tissue slices. Fred's thesis is also very chemically oriented illustrating that at that time biochemistry was still mainly a chemical approach to a subject of biological interest – the emphasis was still on chemistry. His thesis was examined by Charles Harington and the newly appointed Professor of Biochemistry at Cambridge, Professor Chibnall, who had succeeded Hopkins. Harington was not impressed at the large number of spelling mistakes in the thesis. Fred's excuse was that he had typed it himself and was not that good a typist. But he passed when the spelling mistakes were corrected. This emphasis on chemistry in Fred's thesis was to be the mainstay of his next project – the sequencing of the protein insulin.

Fred's introduction to insulin was not his own idea but was stimulated by two events, both entirely out of his own control. First, his supervisor, Neuberger, left for a new job at the National Institute for Medical Research in London so it would have been hard for Fred to continue the work on lysine metabolism in Cambridge, even though that would have been the most logical thing to do from a scientific point of view. Second, Fred needed a job. Fortunately Professor Chibnall had arrived in 1944 on Hopkins' retirement as the new Professor and offered him a position to study proteins, which was Chibnall's main interest. This was the first time Fred was gainfully employed because during his PhD he had lived off the inheritance from his parents. Chibnall's group had made detailed studies on the amino acid composition of bovine insulin. Chibnall encouraged Fred to study the end-groups of insulin because they had already discovered insulin had an excess of 'free' α-amino end-groups. This meant that insulin probably had free α-amino groups at the end of rather short chains. Insulin – a hormone secreted by the pancreas – was a protein of great medical interest because of the discovery by Banting and Best in the early 1920s[8] that insulin could be used to treat diabetes, at that time a death sentence. Perhaps the fact that Fred was himself the son of a doctor attracted him to a project of medical interest.

Fred had not studied proteins during his PhD, although he had worked with lysine and other amino acids present as constituents of proteins, so this was a change in direction for him. However, Hopkins had been interested in proteins such as albumin,[9] so Fred would certainly have known at that time that proteins could be isolated in a pure form and some, including insulin, had already been crystallised by 1926,[10] confirming their purity. The molecular weight of insulin had been overestimated at the time, so there was a great deal of uncertainty about its molecular weight, and it was thought there might be four chains. But this did not deter Fred from starting to study insulin. Many other scientists shied away from this project and thought that sequencing insulin was an impossible problem. There was no methodology to solve it. In fact it was not until about 1950 that Fred himself 'considered that it might be possible to determine the complete peptide sequence of these chains. Up to now both fractions A and B have been investigated but neither is completed.'[11] Up to that point he had simply been looking at amino acids at one end of the insulin chains and some internal peptides. Fred's key contribution was to work out a method of totally sequencing insulin. Temperamentally he had the persistence and imagination to overcome the many hurdles to complete the project. It was certainly not easy.

Fred took ten years from 1944 to 1954 to sequence the 51 amino acids in the two chains of insulin by methods he devised himself. It was known before he started that proteins were formed from chains of amino acids joined together by a 'peptide bond' linking each amino acid. But no one had any idea how the amino acids were arranged in these chains. Was there a regular pattern of amino acids such as a repeating sequence, or not? Was the amino acid chain in a ring forming a circle? Or was there a unique linear sequence with separate N-terminal and C-terminal ends? Fred's main approach was to devise an N-terminal tagging approach by reacting the α-amino end-group of insulin with a fluorodinitrobenzene (FDNB), a simple yellow-coloured derivative of benzene. This yellow tag was stable to acid hydrolysis, which released the end amino acid as a yellow-coloured dinitrobenzene

(DNB) amino acid. The 16 different tagged DNB-amino acids of insulin were identified by their yellow colour.[12] They could be separated from one another and identified, by comparison with known DNP-amino acids that Fred synthesised, using the newly described methods of partition chromatography and paper chromatography of Martin and Synge.[13] By isolating the products of partial acid hydrolysis of tagged insulin and later the products of enzymatic digestion with trypsin, chymotrypsin and pepsin, Fred and his co-workers, including Hans Tuppy, Ted Thompson, Leslie Smith and Mike Naughton, were eventually able to piece together a unique sequence of the two chains linked together by S—S bonds between cysteine residues in the same way that one pieces a jigsaw together. The scientific community was amazed at this tour de force. Moreover this result proved beyond a shadow of doubt that proteins had unique sequences. This contradicted theories that had been advanced before,[14] that proteins consist of periodic arrangements of amino acids, or they were heterogeneous mixtures. Fred stated in the interview that 'proteins were real chemicals with a defined sequence. I mean we always believed this, but it had never been shown, and there were sceptics who used to say that probably proteins were heterogeneous mixtures.' Fred then wanted to know how the insulin amino acid sequence defined its function as a hormone controlling glucose metabolism in the body. But he could only make limited progress on this problem. Nevertheless, the Nobel Committee awarded Sanger the Nobel Prize in Chemistry in 1958 at the relatively young age of 40.

The importance of Fred's work was immense. It stimulated studies of other proteins and many were soon sequenced. Early successes included the amino acid sequence of an enzyme, pancreatic ribonuclease – 124 amino acid residues long, completed in 1960.[15] Soon afterwards, in 1964,[16] the sequence of the proteolytic enzymes trypsin and chymotrypsin, which are responsible for digesting proteins in food in the intestine of mammals, became known. The amino acid sequence of the α and β chains of haemoglobin (the protein that carries oxygen in the blood) and the single-chain myoglobin (the oxygen carrier in muscle) also emerged in the early 1960s.

These advances in knowledge of the amino acid sequence of proteins coincided with advances in X-ray crystallography in Cambridge by John Kendrew and Max Perutz that allowed the detailed three-dimensional structures of proteins to be established. The position of all the main atoms – the carbon, nitrogen, sulphur, oxygen and hydrogen atoms – of the amino acids in space was now known. These structural studies gave new insights into how these proteins functioned as enzymes, how they bound other associated molecules, such as the haem in haemoglobin, and how proteins could change their shape when bound to different molecules.

Ironically, the three-dimensional structure of insulin – the first protein to be sequenced – proved one of the hardest proteins to solve in three dimensions. Dorothy Hodgkin and her colleagues eventually succeeded in 1969[17] more than 40 years after insulin had first been crystallised. Shortly afterwards Fred visited Dorothy Hodgkin in Oxford and is reported to have said on viewing the structure: 'Ah, I got the disulphides right.' Georgina Ferry, in her biography of Dorothy Hodgkin,[18] interpreted this remark to indicate there was a gulf between biochemists and crystallographers, but I doubt that. Fred was fully aware of the power of three-dimensional structures. He, I believe, was relieved, although not surprised, that he had got the disulphides right, since technically this was one of the hardest parts that Fred had to solve in the sequence, because two of the three disulphides in insulin were attached to adjacent residues of the insulin A chain. Sanger wrote in 1988: 'The arrangement of the disulphide bridges . . . proved particularly difficult and probably involved as many man-hours, and more frustrations, than the rest of the work put together'.[19] Although the three-dimensional structure of insulin was now known as a result of Dorothy Hodgkin's work it was still unclear how insulin functioned in the body to control glucose metabolism. So Hodgkin and her team had the same frustrations that Fred himself had found in 1955, when they were wondering which amino acid residues in insulin were important for its function. Now we know that insulin initially binds to specific insulin receptors which belong to a class of transmembrane proteins that cross the plasma membrane of the cell and transmit a signal allowing the receptor to become phosphorylated on tyrosine residues on its internal cytoplasmic domain. Through a subsequent complex chain of signalling events, glucose uptake, as well as fat metabolism, of many cells in the body is regulated. Thus the critical parts of insulin are those that bind to the specific insulin transmembrane receptors and activate its correct autophosphorylation.[20]

After such a momentous discovery and the award of the Nobel Prize in 1958 it might be thought that Fred would want to take up some other activity such as heading a department or going into scientific administration. But he was still a young man of 40, keen to continue to work at the lab bench himself. And he was not interested in running a department or in positions involving scientific administration and committee work. He didn't really think he was any good at lecturing, either. 'So, in fact, it was a stimulus to carry on,' Fred noted. It was now easier to get support to continue his research and he could attract co-workers interested in research. Every head of department wants a Nobel Laureate in their midst because of the prestige it gives. Yet, surprisingly, relations with Professor F. G. Young, who had succeeded Chibnall as head of the Biochemistry Department in Cambridge, were strained at times, probably because Fred and his group, unlike the university-appointed staff, had no formal teaching duties in the department.[21] Fred notes: 'We had nothing to do with the teaching, so we were the poor relations in the Department.' Fred also wanted more space for his research group but this was not available in the department. The Biochemistry Department had to wait almost 50 years until the Sanger Building was built, before it could expand under Tom Blundell's chairmanship. Fred Sanger was invited to open the new building in 1997, saying 'when I was in Biochemistry in the 1950s the lab was somewhat overcrowded but this did not worry us as we were told that we should soon have a new building'.

3

Radioactive sequencing of proteins and nucleic acids

The lean years

Fred described 1955–64 as his 'lean years'[1] as he published relatively little then. In fact, this was a transitional period in which Fred was exploring completely new ideas for sequencing proteins. His idea was to radioactively label proteins.[2] After radioactively labelling a protein and then isolating a radioactive peptide, he would work out the sequence around the labelled amino acid, indirectly. For example, he would use various modifications of the peptide, e.g. removal of the N-terminal amino acid, and monitor changes in the mobility of the peptide on paper chromatography or paper electrophoresis. He was using *the position* of the peptide as a way of sequencing.

This new way of thinking would, in principle, have many advantages. It would not require as much protein as his classical procedures worked out on insulin, since radioactively labelled peptides would be easily and quickly detected on paper by the then relatively new and sensitive method of autoradiography – using an X-ray film. The protein would not necessarily have to be pure, as long as it was radiochemically pure. For example in 1958 Fred incubated an isolated oviduct from a hen with 1 mCi ^{32}P-labelled inorganic phosphate. He successfully isolated ^{32}P-labelled ovalbumin, which was known then to contain two serine phosphates. Fred correctly deduced the presence of one phosphate-containing dipeptide, as serine phosphate-alanine,[3] an observation confirmed ten years later by classical methods of peptide sequencing.[4] Twenty years later on in 1978, in a very different era of DNA sequencing, my colleagues and I showed the exact position of this dipeptide in ovalbumin, as deduced from the nucleotide sequence of a cDNA clone of ovalbumin mRNA.[5]

In fact, Fred had used radioactivity nearly ten years earlier, in 1949, to tag specific amino acid residues in insulin.[6] Then he had chemically modified the tyrosine residues of insulin with radioactive iodine, forming radiolabelled diiodotyrosine peptides. But he decided not to pursue this approach further at that time because of a lack of a quantitative reaction of iodine with tyrosine. But this early use of radioactive tagging of the amino acid tyrosine in insulin showed that Fred was thinking about the use of radioactivity as early as 1949.

But the most useful of the radioactive tagging methods developed in these 'lean years' was the use of ^{32}P-labelled DFP (diisopropylfluorophosphate) – a reactive compound that modified a serine residue in the 'active centre' of trypsin, chymotrypsin, elastase and related serine proteases.[7] This enabled Fred, Brian Hartley and colleagues to show that these proteases were related to one another in sequence, and to confirm five or so residues of sequence around the active centre, entirely by following the labelled peptide or degradation products derived from it. This was no mean feat and highly original research. Fred wrote later: 'The main rationale for this approach was that the radioactive peptides were usually contaminated with non-radioactive substances, but they were radiochemically pure and autoradiographs gave a simpler and more interpretable fingerprint.'[8]

However, Fred had reached the limits of radioactive sequencing of proteins and it was time to turn to nucleic acids. There were just too many possibilities, with up to 20 amino acids in proteins, to deduce a sequence by radioactive methods alone. Nucleic acids were, in this respect, simpler. There were only four nucleotides to contend with.

DNA → mRNA → protein

Nucleic acids had been discovered in 1869 by Miescher[9] but their function was unknown for many years until they were implicated in transferring genetic information in experiments with pneumococci (bacteria causing pneumonia) in 1944.[10] These experiments strongly suggested DNA contained the genetic information. The function of DNA was firmly brought into the spotlight again in 1953 when its double-helical structure was proposed.[11] Watson and Crick concluded their paper, 'It has not escaped our notice that the specific pairing (of bases) we have postulated immediately suggests a possible copying mechanism for the genetic material.' The function of DNA – the molecule coding for the information transmitted from generation to generation – was never seriously questioned after this discovery.

Soon after Fred's description of the insulin sequence in 1955, Vernon Ingram proved that a mutation in a gene caused a change in a protein sequence.[12] This discovery arose from studies of abnormal haemoglobin molecules that caused sickle-cell anaemia – a common inherited blood disorder in Mediterranean countries. Using Fred Sanger's methods, Vernon Ingram found that a particular type of abnormal haemoglobin resulted from a single change of a glutamic acid to a valine. There was now no possible doubt that DNA encoded for amino acid sequences in proteins.

Fred met Vernon Ingram in Cambridge in the 1950s when he was working with Max Perutz's group at the Cavendish Laboratories in Cambridge, where Watson and Crick had made their famous discovery. Informal discussion meetings were held in Francis Crick's house in the evenings. Fred and other members of his research group from Biochemistry attended, along with Sydney Brenner and Max Perutz from the Cavendish Laboratory. The purpose of these meetings was to discuss how genetic problems that Francis and Sydney were interested in could be solved by Fred Sanger's methods of sequencing proteins. Scientists, such as Vernon Ingram, were asked to talk about their current research at these meetings. As Fred's research group grew in size in the 1950s, there was a need for informal meetings of his group, 'Bull Sessions', where one member of the group presented their latest research findings.[13] Later these 'Bull Sessions' continued throughout the 1960s and 1970s as informal meetings held in the evenings in King's College.

The question now was what was the nature of the genetic code? And how did the process of transferring the information from DNA (the gene) to proteins occur? Essentially, the problem was how the information contained in a linear order of four bases – adenine (A), cytosine (C), guanine (G) and thymine (T) – in DNA was 'translated' into a linear order of 20 amino acids in proteins. Evidence for mRNA (messenger RNA), the key intermediary molecule between DNA and protein, was discovered around 1960.[14] Crick summarised this work in his famous central dogma 'DNA→ mRNA → protein', showing the flow of information from DNA via mRNA to protein.

The genetic code was eventually solved around 1965 by indirect methods, and turned out to be a triplet code. That is, a given amino acid was specified by three nucleotides in mRNA. Thus, for example, UUU in mRNA coded for the amino acid phenylalanine. (U codes for uracil in mRNA, replacing thymine in DNA.) The code was non-overlapping, meaning that the linear mRNA was read strictly in sequence, i.e. UUU,AAA codes for the amino acid sequence phenylalanine lysine. We shall see that Fred Sanger confirmed this genetic code in his later RNA sequencing work.

The biochemical process by which the information in DNA is converted into a protein sequence involves essentially two steps. The first is the copying (or 'transcription') of the DNA by an enzyme, RNA polymerase, to synthesise the mRNA. This is followed by the reading of the linear order of bases in mRNA by peptidyl synthesis (translation) catalysed by ribosomes (intracellular organelles where protein synthesis occurs).[15] Another nucleic acid turned out to be key in this process and became known as tRNA (transfer RNA),[16] there being transfer RNAs specific for each of the 20 amino acids – the 'adaptors'[17] that recognised both amino acids and the mRNA. The ribosomes also contained RNA – called ribosomal RNA. Both tRNA and ribosomal RNA were relatively stable RNA molecules differing from the unstable, or rapidly turning over, mRNA.

Early DNA sequencing

Nucleic acid sequencing had been attempted long before Fred got involved. Ken Burton and George Petersen, working in Oxford in the late 1950s, perfected the 'depurination reaction'. This degraded DNA into very short 'runs' of pyrimidine residues, that is, thymine (T) and cytosine (C) residues. By this means they could separate, by column chromatography, and sequence for example all three isomers of the trinucleotide (C_2T), i.e. CTC, TCC and CCT.[18] Later, in 1965, Murray and Offord,[19] working in Fred Sanger's laboratory, also used enzymatic digests of DNA to isolate and characterise short DNA oligomers using the two-dimensional methods that Fred had introduced for his RNA sequencing work. By using a combination of radioactive ^{32}P-labelled DNA that was isolated from the bacterium *E. coli*, mixed with unlabelled calf thymus DNA that was post-labelled by neutrons in a specialist reactor, they showed there were clear differences in the frequency of quite short DNA sequences. Although ingenious, this technique could not easily be pursued further because of the long timescale of the experiment and the need for access to a powerful external neutron source. The real problem, however, in the 1950s and 1960s, was that there was no short DNA molecule available to test sequencing methods on, whereas there was the hope that one could fully sequence the tRNAs that had been recently discovered. These tRNAs were of reasonably low molecular weight. So Fred started to sequence RNA. DNA had to wait.

The new Laboratory of Molecular Biology

Fred started RNA sequencing at the time he moved his research group from the Department of Biochemistry to a new purpose-built laboratory on the outskirts of Cambridge. Space had been found for the new building, sponsored by the Medical Research Council (MRC). This Research Council had supported Fred since 1952, and others within his group, such as Brian Hartley and Ieuan Harris. Fred's successful research, despite his 'lean years', had been critical, along with that in physics of Francis Crick and Max Perutz at the Cavendish Laboratory, who were also MRC-supported, in persuading the MRC to establish the new Laboratory of Molecular Biology in Cambridge.

It probably helped that Fred was awarded the Nobel Prize in Chemistry at a critical juncture in 1958 when Max Perutz was negotiating with the MRC to join Sanger's and his own group in a single new building. Knowing they would have Francis Crick, Jim Watson (although Watson never actually joined the new lab) and now Fred Sanger in a new laboratory was obviously attractive to the MRC. In retrospect it was a very bold move. The plan was to join together biophysicists with a limited interest in biology, with biochemists with a limited interest in physics. It might not have worked. Credit must be given to Harold Himsworth, 'The Secretary' at the MRC, and Max Perutz, who was very persuasive on behalf of the scientists, for making it possible.[20] Geneticist Sydney Brenner, and another crystallographic group headed by Aaron Klug, also joined the new laboratory. The new lab at Max Perutz's insistence had no Director – only a chairman, himself! To begin with it was run by a group of four: Max Perutz, John Kendrew, Francis Crick and Fred Sanger.

The lack of support from Cambridge University for alternative arrangements to set a up a laboratory of molecular biology in the *centre* of Cambridge, close to the existing science laboratories, reflected the different pressures on Professor F. G. Young, as Professor of Biochemistry, and Professor Mott, at Physics, since they had to ensure teaching of undergraduates, whereas Sanger, Perutz and Crick wanted full-time support for their research without the distraction of university and college teaching. Apparently there was no obvious new space that Young or Mott could come up with within the central university area to satisfy the ambitions of Perutz and Sanger. Professor Young was particularly critical of the plan to set up a research laboratory,[21] arguing it would be a risky move and such 'Research Institutes' did not succeed in the long run. How wrong Young turned out to be. But it was agreed that PhD students who had registered at Cambridge University could be affiliated to the new Laboratory of Molecular Biology – a requirement emphasised by John Kendrew. This decision turned out to be critical and allowed me and many other PhD students to benefit from the exposure to scientists, like Fred Sanger, who were wholly committed to research. The new Laboratory was to be over a mile out of town on the New Addenbrooke's Hospital site and also housed a new University Department of Radiotherapy,[22] opening in 1962.

RNA sequencing

The move to start sequencing RNA was, as Fred says, 'quite obvious, I suppose. I don't think there was any sudden decision to work on RNA. You work on something you think you can solve. I think before (the discovery of tRNAs) . . . there was no real possibility of working on nucleic acids.'

Transfer RNAs (tRNAs) were relatively short nucleic acids, about 80 nucleotides long, amenable to having their sequence determined, although they posed an initial difficulty in that they occurred as a mixture of 20 or more closely related molecules, so it was non-trivial to purify one particular tRNA. But the decision to work on RNA was also gradual, and initially a tentative decision, since Fred was unsure how well his new methods would work. Unlike his competitors at the time,[23] Fred wanted to develop radiolabelled methods. His idea was to introduce a labelled ^{32}P into the phosphate backbone of the nucleic acid as a tag. He would then use this radiolabel to detect the nucleic acid, or fragments derived from it, rather than the classical way of detecting nucleotides by their absorption under ultraviolet light. He predicted, correctly as it turned out, that this methodology would be quicker and easier in the long run for sequencing. He was the first to introduce ^{32}P into RNA chemistry and it was a highly innovative approach, seen as risky by some commentators at the time. But Fred was not easily put off. Moreover, there were highly specific nucleases available, such as the guanine (G) specific T1 ribonuclease, to degrade RNA into small fragments.

Fred tried out his radioactive sequencing methods by adding ^{32}P inorganic phosphate to a growing culture of *E. coli* or yeast. He then allowed the growing cultures to synthesise uniformly ^{32}P-labelled nucleic acids. After isolation and purification of RNA, initially ribosomal RNA, he attempted to separate the complex mixture of T1 ribonuclease fragments by two-dimensional separations on modified papers. This proved difficult at first but he described his first real success when his recently appointed assistant, Bart Barrell, brought a beautiful autoradiograph, with clear round spots, to him after an X-ray film had been exposed to the modified paper overnight. (Fred always called them radioautographs, rather than autoradiographs!)

Fred was very keen to establish the sequence of the spots on the radioautograph from 'positional' information, i.e. where the compounds were present on the radioautograph.

> The thing about this is that we were very keen – as we were with the radioactive proteins – to try to find methods in which you identified sequence from a *position* on a two-dimensional system, or position on a one-dimensional system. In fact you can on this system, for small

oligonucleotides, because there are only a limited number of nucleotides. We could identify them from position on the two-dimensional system.

Fred did not complete the sequence of a tRNA, phenylalanine tRNA, until 1969.[24] He had been beaten to sequencing the first tRNAs by Bob Holley who described the sequence of alanine tRNA from yeast in 1965, for which Holley was awarded the Nobel Prize in 1968. In fact, the first success from the Sanger lab with short RNAs was with another small RNA, the 5S ribosomal RNA, in 1967, the subject of the author's PhD with Fred Sanger.[25] I had been asked by Fred to purify phenylalanine tRNA when I started my PhD, but had failed to make progress in nine months' work. So I suggested, instead, that I purify a small RNA that had just been described, and which subsequently turned out to be a small 5S ribosomal RNA. This molecule was a useful model compound to test out Sanger's methods of radioactive sequencing, since it was relatively easy to prepare in good radioactive yield in a pure form. We described its sequence in 1967. It was 120 nucleotides long and for a short time was the longest known RNA sequence, proving the validity of the radioactive methods Sanger had pioneered. A particular problem in sequencing this RNA was the problem of separating longer partial T1 ribonuclease digestion products from one another to obtain longer fragments. This required the development of another, new, fractionation system called 'homochromatography' that was a type of displacement chromatography on ion-exchange paper and later adapted for use on thin layers.[26] Again, this development was typical of Fred's ingenuity in devising completely new methods. It was vital for the 5S sequence but became even more important in later work on R17 bacteriophage RNA sequencing and in the development of DNA sequencing.

Fred next wanted the challenge of sequencing longer RNA such as the bacteriophage, 'phage' R17 RNA, but this was a formidable challenge in the 1960s.

Fred records: 'That sounded quite a big jump, really. It seemed a bit impossible. It was 3000 residues long – it's a bacterial virus and one didn't think it would be possible to simply partially hydrolyse it and get fragments out because of its great size. You would think you would get a hopeless mixture of large fragments if you did a partial T1 digest.'

But to Fred's surprise Jerry Adams, a visiting postdoc in Fred's laboratory, did manage to get out discrete RNA fragments of this bacteriophage RNA. He used polyacrylamide gel technology to separate out partial T1 ribonuclease fragments of ^{32}P-labelled R17 RNA. This R17 RNA formed the RNA genome of this phage, but also functioned as its mRNA. One partial fragment proved particularly interesting, since it coded for part of the coat protein of this phage that had been previously sequenced by protein sequencing methods. Thus Fred and his colleagues were able to align the amino acid sequence with the nucleotide sequence and obtain, for the first time, the sequence of 67 nucleotides of mRNA directly.[27] It confirmed the genetic code. The scientific community was not surprised at the time that the genetic code was correct, but Fred records:

> That was quite exciting because that was the first time that a nucleotide sequence had been determined and shown to be related by the genetic code to a known amino acid sequence in a protein. I mean one of the purposes of going into nucleic acid sequencing was to try and break the code, but in fact the code had already been broken by the time we got there. We weren't the first people to break the code but it was a good confirmation of the code. I think it was worthwhile.

DNA sequencing

Fred and his colleagues never completed the sequence of R17 RNA. Fred left that to the Fiers lab, who completed the sequence of a closely related MS2 RNA in 1976.[28] But DNA was now in his sight. The basic problem was summarised by Fred:

> The problem with DNA at that time was the obvious limitation that there was no way of attacking it because of its very large size. The simplest DNA known then were the single-stranded DNA bacteriophages. The one that featured largely in our work was the φX174 DNA, which had about 5000 nucleotides. It was obviously a bit big to try and start doing sequences on, so it didn't look as though we could apply some of our partial hydrolysis methods to that DNA especially as there weren't suitable enzymes to degrade DNA nice and cleanly – like the T1 RNase that we had used, which was very useful for the RNA work. There wasn't anything like that in the DNA field.

Nevertheless, significant progress was made with partial digestion of bacteriophage φX174 DNA with an endonuclease IV. John Sedat came in 1970 as a postdoc from Sinsheimer's lab in California bringing with him the bacteriophage φX174 which he was familiar with. Under the right conditions this enzyme produced reasonably specific degradation products of ^{32}P-labelled φX174 DNA on polyacrylamide gels – a separation method similar to the one used by Jerry Adams earlier on R17 RNA. Early success emerged with the sequence of 48 nucleotides of φX174 DNA in 1973, in collaboration with Ed Ziff and Francis Galibert in a paper entirely produced in Fred's lab[29] – but without his name as an author. John Sedat had expected Fred to be an author on this paper. His name had been included in the draft manuscript with Ed Ziff and Francis Galibert, but Fred declined authorship.[30] Fred took this approach quite often in his career, as he felt he had to make a significant contribution to a piece of work for his name to be included. It was not enough that he was the head of the laboratory in which the postdoc or PhD student had done the work.

Another important result from Fred's laboratory about this time was the work of Vic Ling in 1972.[31] He sequenced some of the longer depurination products[32] of another relatively small single-stranded bacteriophage, fd DNA, yet Fred was not an author. This paper convincingly showed it was possible to deduce the composition of the depurination products of this phage fd DNA simply from their position on the two-dimensional fractionation system. Admittedly, the fact that the degradation products contained only T's and C's simplified the pattern of spots, and thus the deduction of composition from position. Positional information was used twice in this study, not only to deduce the composition of nucleotides from their position, but also in simplifying the method by which sequence of some of the longer products was deduced by partial exonuclease digestion. This method – really an extension of methods previously used in sequencing RNA – used a two-dimensional fractionation system on thin layers to analyse the complex mixture of products and deduce their sequence simply by inspecting the relative position of successive degradation products. Remarkably Vic Ling deduced the sequence of a unique 20-long fd DNA fragment – quite an achievement at that time. This theme of using positional information on a given separation method to simplify the method of deducing sequences was a common theme of Fred's RNA and DNA work. We shall see that this principle of using the relative position of nucleotides on separation systems eventually led him to his definitive method of sequencing DNA in 1977.

In the early 1970s, other ideas of how to sequence DNA were emerging. It might not be necessary to uniformly label the DNA of interest with ^{32}P by growing the phage in the presence of ^{32}P-phosphate. Why couldn't the ^{32}P tag be introduced into the DNA after its isolation? This might suggest more specific ways of interrogating the DNA in small sections, rather than trying to sequence the DNA as a whole.

Fred recalls the first such successful approach that involved copying DNA with an enzyme. This was the method used to sequence 12 nucleotides at the end of another bacteriophage DNA, λ DNA, in the early 1970s.

> We were not the first people to use copying. The first people to copy DNA were Wu (Ray Wu) and Kaiser, and that was a classical piece of work. They were studying the 'sticky' ends of phage λ and they used DNA polymerase. Now DNA polymerase requires a single-stranded DNA template and a primer, which will hybridise to a position on the template, in other words, a piece of single-stranded DNA followed by a piece of double-stranded DNA. Then it will copy the DNA template from the 3′ end of the primer. They actually sequenced the end of the λ by quite complicated methods. I think it took them about a year to do it. But that was essentially the method we followed. But the method as described was only applicable to the sticky ends of λ – these 12 nucleotides.[33] We wanted to try and develop a general method whereby we could use the copying by DNA polymerase to really get a short radioactive piece of DNA. If you take a primer[34] on a single-stranded DNA of a phage, then you could make a short piece of radioactive DNA.

Thus in the early 1970s Fred had started to take an interest in methods of sequencing DNA that involved synthesising ^{32}P-labelled DNA, or RNA, using copying enzymes – polymerases, and radiolabelled nucleoside triphosphates as substrates for these enzymes. Instead of introducing the radiophosphate label by growing the phage in the presence of ^{32}P-phosphate, he started with unlabelled DNA and introduced the labelled ^{32}P-phosphate via an enzymatic reaction with DNA or RNA polymerase. Experiments by Fred's PhD student, Elizabeth Blackburn (later to become a Nobel Laureate herself in 2009 because of her pioneering work on chromosome ends), in which a short segment of φX174 was copied by RNA polymerase were promising,[35] but the better idea was to copy with DNA polymerase. This was preferable because only that region of DNA to which a short oligonucleotide 'primer' had hybridised would be copied. In that way one 'sampled' a very *short* section of a very *long* DNA. This new idea, however, required an oligonucleotide primer, which was very difficult to synthesise in the early 1970s. Gobind Khorana in the USA had developed the methodology for oligonucleotide synthesis but it still took an expert chemist a year to synthesise one.

Fred recalls:

> But to make that . . . oligonucleotide primer . . . at that time was difficult because nucleotide synthesis was extremely difficult. It was being pioneered largely in the lab of Khorana (Massachusetts Institute of Technology, MIT) but it was extremely tedious. At a conference I met Hans Kossel, who had worked with Khorana, and he had a similar idea to what we had – to make an oligonucleotide having a sequence complementary to some DNA sequence of a phage and to extend it and make it radioactive. So we decided to get together. He had all the experience of making oligonucleotides, and he made two oligonucleotides and it took them over a year to synthesise these eight residues in each oligonucleotide. But we received them and that was a basis for the next stage of our work.

On receiving the oligonucleotide primers from Fischer and Kossel, Fred working with Alan Coulson (his assistant) and with John Donelson (a postdoc from the USA) copied the bacteriophage f1 DNA with DNA polymerase. One of the oligonucleotides used in this copying reaction was an 8-long primer, that was designed to hybridise to a predicted f1 DNA sequence, coding for the sequence of the tripeptide Try.Met.Val of its coat protein. This primer was chosen for the very good reason that this particular amino acid sequence allowed one to deduce a unique primer sequence, because the codons for tryptophan and methionine are unique in having a single, rather than one or more alternative codons, because of the redundancy of the genetic code.[36] Ironically the primer did not behave to order. It did not hybridise where it should have. Luckily it hybridised elsewhere and specifically. This setback might have concerned many scientists, but did not put Fred off in the slightest. He set out to synthesise radioactively labelled DNA products using his own in-house synthesised ^{32}P-labelled deoxynucleoside triphosphates, helped by the expertise of Bob Symonds, a sabbatical visitor from Adelaide, Australia.

But there was another problem. How was one to analyse these DNA products? Fred was not keen on endonuclease IV, in use by other colleagues in his lab,[37] as it lacked any absolute base specificity. He chose instead to investigate a known property of DNA polymerase that, under specialist conditions, 'misincorporates' *ribo*nucleotides instead of *deoxy*ribonucleotides into the synthetic product. Thus the product was really a hybrid of DNA and RNA; and Fred arranged that every C residue incorporated into the synthetic DNA was RNA in origin. This enabled him to degrade the hybrid DNA–RNA product using the highly specific pancreatic RNase that cut only at the C residues, leaving the DNA part of the product intact. He was then able to deduce the sequence of 50 residues of DNA. This was a significant achievement at the time,[38] but it was still not very satisfactory, because it involved the time-consuming method of eluting many individual spots from the two-dimensional separation method and analysing these spots further. But the method did show that DNA copying methods were capable of producing a specific and presumably correct sequence. If errors in copying occurred they were at too low a rate to influence the results.

But an even more important idea emerged in this work. It was noted that products accumulated on the two-dimensional separation method, immediately *before* the radioactive nucleotides were incorporated – presumably because of the lower concentration of the radioactive substrate compared to the non-radioactive ones.[39] This suggested that it might be possible to devise a sequencing procedure *in one dimension* based simply on the position of spots. Fred spent quite lot of time trying to get a reasonably random set of labelled products synthesised by the DNA polymerase ending at a position known to require the addition of a particular residue – say an A residue. As a back-up procedure, Fred used another method with a different, T4 DNA polymerase known to synthesise products ending in (say) an A residue.[40] Combining data from each of the other three nucleotides (C, G, T) with data from the products ending in A should give one a method, in principle, of sequencing in one dimension without the time-consuming work of eluting spots and analysing them further. This method became known as the 'plus and minus' method.[41]

According to Fred, John Donelson performed an experiment that was crucial for this new method. He used polyacrylamide gels, in one dimension, to separate the samples Fred himself had prepared. The results on this new polyacrylamide gel system were compared with homochromatography, the standard separation method used for larger oligonucleotides up to that point. John Donelson and Fred found that acrylamide gels were better than homochromatography.

Fred recalls:

> At the time we started out John Donelson was doing some experiments with acrylamide gels. We were largely using homochromatography because we thought that was rather better for the larger oligonucleotides, but I gave some of these samples to him and he ran them on acrylamide gels and they did look marginally better than the homochromatography. They weren't very good at the time but they were run on much smaller gels than we would now normally use – yes, smaller gels, and not under denaturing conditions. We then started to use much longer gels under denaturing conditions.

It is now accepted that acrylamide gels are perfectly suitable for separating *individual nucleotides* up to 500 or more long, but then this was just another 'crazy idea' then. No one had thought of separating individual nucleic acids, that differed only by a single residue, on such gels before, although such gels were commonplace at the time for separating intact proteins, RNA and DNA fragments.

Fred recalls:

> The unexpected thing about this is that you could find a method that one could fractionate fragments absolutely according to size. And it depends on that. I think when we first started doing this it was just another crazy idea. How could we possibly find a system that would separate things exactly according to size? So we didn't take it too seriously, and I didn't make careful notes when we started on it. But in fact you can separate things according to size on acrylamide gels. It was not too easy at first. On the first gels we found that we got very bad inversions and sometimes the larger fragments were running faster than the small ones. And that completely upset things.

So the plus and minus method was developed based on separation of bands on one-dimensional gels. There was now no longer any need to elute bands or analyse them further. One simply 'read off' the sequence from the relative positions of bands in adjacent lanes of the radio-autograph. Fred has stated: 'This new approach to DNA sequencing was I think the best idea I have ever had, being original and ultimately successful.'[42]

This plus and minus method was applied to the sequence of bacteriophage φX174 DNA and its sequence of 5000 or so nucleotides were worked out, with a few uncertainties. It was typical of Fred's cautious approach that he admitted the sequence probably had errors in it.[43] This was the first paper published in the field, as far as this author is aware, that admitted there could be errors in a published sequence.

To many people's surprise Fred had yet another good idea up his sleeve. The plus and minus method depended on copying methods by DNA polymerases designed to specify which of the four possible terminal residues, either C, A, G or T, was at the end of the synthetic fragments. Another way, in principle, of defining an end was to use a synthetic analogue that differed from the normal substrates, such that the DNA chains could not be extended further. Such 'chain terminator' molecules were dideoxynucleoside triphosphates, or 'dideoxys' for short. One such analogue, dideoxyTTP (ddTTP), had been described before by Arthur Kornberg. Fred found on testing this that he obtained a much more even distribution of bands on the acrylamide gels than with the plus and minus method. He realised this would enable better gel reads with fewer ambiguities, and longer reads might be achieved. The trouble was the other three dideoxys had never been synthesised. Undaunted, Fred and Alan Coulson set out to synthesise them. I doubt that Fred could have done this without his extensive chemical expertise acquired during his PhD thesis work nearly 35 years earlier with Neuberger. Either that, or he was now supremely confident that it would be an important breakthrough if they could synthesise the three remaining dideoxys. And so it proved. After a year's work they had synthesised the other three dideoxys.

Fred recalls:

> The other three dideoxys had never been made and were not available anywhere. So Alan Coulson and I had to set down and just prepare

> them! This was something which we had really no experience of – nucleotide synthesis. Luckily, we had Mike Gait in the lab – an expert in nucleotide synthesis, and Bob Shepherd and they were very helpful to us, and I had a certain amount of experience of chemical synthesis. But nucleotide synthesis is rather specialised. It's not quite such fun as making dyestuffs and things where you get beautiful crystals. You can never crystallise these nucleotides. You just separate them on some type of chromatogram and see how pure they are.

This dideoxy method published in 1977 was the definitive method as far as Fred was concerned.[44] It allowed his group to recheck the sequence of bacteriophage φX174 DNA and correct the 30 remaining errors in the 5386-long sequence.[45]

But there was still another problem to be overcome. Fred had always selected single-stranded phage DNA to sequence, avoiding the problem that almost all DNA is double-stranded. He therefore set himself the problem of sequencing the 49 000 or so nucleotide sequence of the double-stranded phage λ DNA sequence, which he completed with the help of colleagues in 1982.[46] He did this by first cloning the double-stranded sections of phage λ into a single-stranded phage, bacteriophage M13, thereby converting double-stranded DNA to single-stranded DNA suitable for his dideoxy sequencing method.

Overlapping genes

No one now doubted that DNA sequencing had come of age, and even longer DNAs might now be sequenced, including the human genome. The shortest known human DNA was mitochondrial DNA which turned out to be 16 500 or so bases long, sequenced in 1981.[47] This was the beginning of the Human Genome Project and was the first substantial piece of human DNA to be sequenced. There was now talk of sequencing longer genomes, such as the *E. coli* genome, and for the first time the entire human genome. Fred recalled:

> I think . . . the human genome project . . . is a logical conclusion of the work I've been doing really. I think 50 years ago I felt sequences were the important thing. I have spent most of my life trying to develop methods for sequencing. And of course the DNA sequence is the ultimate – you

can get the sequence of the RNA and protein from the DNA. So I feel that it is a very worthwhile project. And from the medical point of view it is likely to prove useful.

Although Fred emphasised that his main interest was in developing the methods, to his and everyone else's surprise, sequence features emerged from his method-directed approach that no one had expected.

The first surprise was that genes could 'overlap' with one another in bacteriophage φX174 DNA.[48] Biologists and geneticists previously believed there were no exceptions to 'the one gene, one enzyme' hypothesis,[49] but now a complication arose in the sequence of φX174. It was found that one region of this DNA codes for the amino acid sequence of not one, but two proteins. The way this was achieved was by using different 'reading frames' of the triplet genetic code. The first protein was coded by the *first* reading frame of the genetic code and the second protein by the *second*, overlapping reading frame. Thus one region of DNA coded for two quite different proteins. In other words the 'one gene, one enzyme' hypothesis of Beadle and Tatum did not apply in this case.

Different genetic code in mitochondria

Another surprise was that the genetic code was not universal. This code had been worked out in the mid-1960s by indirect methods and Fred's work had amply confirmed this code. So it was quite a surprise when differences were found in mitochondrial DNA.

Fred recalled:

> We found a lot of sequences which appeared to be coding for proteins but contained terminating codons – UGA. And this was a bit of a mystery, because this is a terminator. You couldn't have a protein with UGA in it, because the genetic code was considered to be universal, and UGA was a terminator! And this was a big puzzle but eventually when we put a whole lot of sequence together and using amino acid sequence, as well as nucleotide sequence, it was clear that this UGA was *not* a terminator. It was actually a codon for tryptophan. This was the first time that the genetic code, which was thought to be universal, was shown not to be universal. Mitochondria had a different genetic code from the normal genetic code used in most bacteria and other

> mammalian organisms. And, we found a few other differences in mitochondria,[50] and differences between the yeast and human mitochondria.[51]

Fred Sanger's assistants: Bart Barrell and Alan Coulson

Fred delegated some of the human mitochondrial work to his trusted and long-term assistant, Bart Barrell. Indeed Fred was not an author on perhaps two of the most interesting results paper from his group on overlapping genes and the altered genetic code of mitochondria – papers on which Bart was the senior author.[52] Bart had come to work with Fred straight from school in 1964, only 18 years old, without first studying at university. Fred had been impressed with his enthusiasm and enterprise and had taken him on. Bart had been involved in all the nucleic acid developments particularly those dealing with tRNA sequences. He was, therefore, particularly good at predicting where the mitochondrial tRNAs were encoded in the mitochondrial DNA sequence, as Fred wrote:

> When we were working on these larger DNAs it was necessary for someone to collect from the various contributors the data to be put into the computer and analyse it. This was usually done by Bart Barrell, who had considerable experience and was an expert at studying sequences and picking out interesting features, such as tRNA genes in mitochondria. He was the first to spot the overlapping genes in φX and the altered code in mitochondria. This was a somewhat sedentary occupation and I myself preferred to be working at the bench, particularly on the development of methods.[53]

Another long-term colleague, Alan Coulson, joined Fred as a technician, when he was 20 years old, in 1967. Like Bart Barrell, he had never been to university although he had completed a Higher National Diploma at Leicester Polytechnic. Alan was an author on the critical paper describing the dideoxy sequencing method for DNA in 1977, and published his last paper with Fred in 1992, 25 years later. He was co-author of 20 papers with Fred – more than any other scientist, so can claim to be Fred's closest colleague. He has a very similar temperament to Fred, is modest about his contributions, and did not seek publicity, although Leicester Polytechnic – now De Montfort University, where he studied as a sixth-form student – recognised his exceptional contributions to science and awarded him an honorary DSc in 1992. More recently his laboratory notebooks have been archived, along with Fred Sanger's own extensive lab notebooks, at the Wellcome Archive in Euston Road, London. This may be a unique honour. Both a double Nobel Prize winner's and his technician's experimental, largely hand-written notebooks, were deemed worth of archiving for the benefit of future historians.

Alan recalls that on his first day of work in his new job Fred said, after quite a short, introductory explanation of his current work, 'Well, we'd better get on with it then.' Fred believed in 'getting on' with the experiments. The reading could come later. In fact there was not much literature to read, as Fred's approach was nearly always highly original.

Alan Coulson admitted to being completely naive as an experimentalist when he joined Fred. Alan had learned about RNA and DNA in his Higher National Diploma course at Leicester Polytechnic but had no idea of what was actually involved in research.[54] Alan never had a desire to become scientifically independent of Fred and establish his own research group, unlike Bart Barrell. He was happy to continue to work as Fred's technical assistant until Fred's retirement in 1983, after which he joined John Sulston. With Sulston, Alan was involved in the genome sequence of the nematode worm *Caenorhabditis elegans* and the Human Genome Project. Alan did, in fact, gain independence early on in helping Uli Rensing and Bart Barrell and others to sequence RNA. These three papers were published independently of Fred Sanger.[55] Alan never gave a talk throughout the entire period he worked with Fred from 1967 to 1983, although he did attend lab seminars and the evening 'Bull Sessions' held in King's College.[56]

Alan knew Fred's pattern of work better than most of Fred's colleagues as he had worked with him over such a long period. Fred usually arrived at about 9.30 in the morning and often worked in the evenings. Fred never took a morning coffee break in the canteen, and only rarely broke for afternoon tea, but invariably took two hours off for lunch. He walked home to have lunch with his wife, Joan, at their house nearby in Hills Road. One story is typical of the way Alan and

Fred worked together. When Fred and Alan were developing the read-off methods of DNA sequencing, Fred had decided to synthesise the three dideoxynucleoside triphosphates that had never been described before. Alan, as Fred's assistant, was closely involved in the new syntheses. Alan told me that a flask containing one of the newly synthesised dideoxy triphosphates, that had been freeze-dried in the flask, was accidentally dropped on the floor by Fred and smashed into lots of pieces. Alan then decided to pick up the pieces of glass from the floor, washed them and recovered the precious material. It actually proved to be the correct product when tested in a dideoxy sequencing reaction later.

Fred was considerate to Alan and to his family – almost a friend of the family.[57] He came to Alan's wedding to Sue at the Shire Hall, Cambridge, in 1971. Sue was also invited by Joan Sanger to the Sanger family home in Hills Road to have tea when Alan's children were quite young. My wife, also Sue, and our children, Edward and Elizabeth, were also invited at the same time. Fred had been sent up into the loft to get his own children's toys down for this new generation of staff's offspring to play with.[58] This concern for the families of his staff showed Fred, and his wife Joan, wanted to support his staff in more ways than simply encouraging their science.

There were other occasions when Fred supported his staff's outside work activities, for example when Fred helped John Sedat[59] shortly after his arrival from Sinsheimer's lab in 1970, to obtain an association with a Cambridge college. Fred arranged this quite quickly with University College, later to be renamed Wolfson College, in 1973. This college association allowed John to invite the only woman scientist working in Fred's group, Liz Blackburn – a young Australian PhD student – to play tennis at the college courts. Their relationship blossomed over the next few years, and they married in 1975 soon after they both took up postdoctoral positions in the USA.

Fred always was careful to maintain good relationships with his colleagues, whether they were Visiting Professors, e.g. Mark Ptashne from Harvard, George Petersen from Dunedin, New Zealand, Ted Friedmann from San Diego, California, or postdocs on prestigious fellowships from the States, or PhD students, or his technicians. Fred thought the most important thing was to be able to get on with his staff as people. That way they were most likely to contribute to a successful research effort. Fred recalled, when considering who to appoint to a postdoc position in his lab:

> It is not only academic brilliance you are looking for. For a successful collaboration one of the most important things is to have someone you like and get on well with, and references are often not helpful in this respect. When I was considering John Donelson, one of the referees has added a postscript saying 'I think you will enjoy his sense of humour.' This proved to be the case.

If there were problems within his group, Fred largely ignored the problem, just waiting until matters resolved. He did not actively try to resolve problems, perhaps realising this would not necessarily help matters. Fred wrote much later, after his retirement, in 1988:

> When we started our work on DNA we had a relatively large number of people and I noticed a certain amount of rivalry and bad feeling seemed to be building up. This worried me rather and I did not know how to deal with it, never having had the problem before, so I largely 'acted ignorant' and took no notice of it. I think this was probably the best policy, by assuming that they were all friends they were shamed into recognising the stupidity of their behaviour. For successful collaborative work there must be complete understanding and trust, and a sharing of ideas and resources. If friction arises usually the work suffers.[60]

Fred had a very well-developed work ethic that can be traced back to his Quaker upbringing and his schooling. Fred worked himself in the lab and was an experimentalist. He was often there in the evenings – not in his office, but working in the lab.[61] Fred expected his colleagues to work hard, as he himself did. Nevertheless Fred could be very sociable at times. Alan Coulson recalled an occasion when Fred brought his own, homemade, elderberry wine into the Christmas party at the lab.[62] Another time, in 1981, when the phage λ sequence was completed Rodger Staden – the computer expert – wrote a new computer program whereby Fred would press a computer key that would activate a ringing tone from the computer. This would be the signal for those present (Fred, Bart Barrell, Alan Coulson and George Petersen) to stop work and celebrate with champagne.[63] There was an earlier occasion when there was to be a prize of three weeks' holiday for the first person to sequence more than 50 bases of DNA. By 1973, ten scientists within

Fred's lab could claim this prize, so all ten took the weekend off instead.[64]

The second Nobel Prize

The second Nobel Prize was awarded to Fred Sanger, Walter (Wally) Gilbert and Paul Berg in 1980. Fred and Wally Gilbert shared half of the Chemistry Prize for sequencing nucleic acids. Paul Berg was awarded the other half for his fundamental studies of the biochemistry of nucleic acids, with particular regard to recombinant DNA.

It is most unusual for anyone to be awarded two Nobel Prizes, and Sanger is only one of four people to have been awarded the Nobel Prize twice. The other three were Marie Curie, first for Physics, along with her husband Pierre Curie and with Anthoine Becquerel for their work on radioactivity in 1903. Later, in 1911, Marie Curie was awarded the Prize in Chemistry for her discovery of polonium and radium. Linus Pauling was awarded the Prize in Chemistry in 1954 for his work on the structure of proteins, in particular of their α and β helical structures, and later in 1962 the Nobel Peace Prize. John Bardeen was awarded the Nobel Prize in Physics, first in 1956 with William Shockley and Walter Brattain for the invention of the transistor; and again in 1972 with Leon N Cooper and John Robert Schrieffer for a fundamental theory of conventional superconductivity.

What was the relative contribution of Sanger and Gilbert to DNA sequencing? Were they equals, as their joint award of half the Nobel Prize suggests? There had been a continuous series of papers from the Sanger lab over a period of 12 years from the early description of fingerprinting RNA in 1965,[65] culminating in the dideoxy DNA sequencing method in 1977. On the other hand, Gilbert published one definitive paper in 1977,[66] describing a novel chemical degradation method for sequencing radioactively labelled DNA. Gilbert most likely had benefitted from Sanger's development of radioactive approaches to sequencing and the development of gel technology. Arguably, Gilbert was fortunate to have been recognised as Sanger's equal by the Nobel Committee. I assume the committee must have taken into account Gilbert's other important contributions to molecular biology.

Fred, commenting on Gilbert's contribution, said 'I cannot pretend that I was overjoyed by the appearance of a competitive method'.[67]

Maxam and Gilbert's method became available about the same time as the dideoxy method in 1977, after the plus and minus method. It was an ingenious, chemical degradation method, although less specific for individual nucleotides than Sanger's methods. Unlike the Sanger method, it could be applied to double-stranded DNA. This initially proved attractive, and the method was widely used especially in the USA, since there was no need to prepare single-stranded DNA. Like Sanger's methods, the Maxam and Gilbert method used 'read-off' of radiochemically labelled bands on acrylamide gels, using similar gel technology. But the Maxam and Gilbert method involved a number of chemical steps and was not as easy to automate as the Sanger dideoxy copying method, which proved the method of choice for the subsequent high-throughput sequencing necessary to complete the nematode worm and human genome projects much later in the 1990s. The discovery of heat-resistant DNA polymerases, which were isolated from bacteria that inhabited hot springs, allowed one of the strands of double-stranded DNA to be copied without the need to isolate single-stranded DNA. This meant that Sanger sequencing could now be easily applied to both double-stranded and single-stranded DNA. The Maxam and Gilbert method became redundant and is no longer used.

Was it easier to get the second Nobel Prize than the first one, was a question I asked Fred.

> I think it was really. When you've got a Nobel Prize you get good facilities for research, you get good students, excellent postdocs who used to come to the lab and work with us. And another thing, having had a Nobel Prize, I didn't feel there was any obligation to get papers published. I could spend my time doing rather crazy 'way-out' experiments. I think this is how you get important advances very often, by trying the way-out experiments that are very likely not to work – crazy experiments. I did lots of experiments that didn't work. I enjoyed it more and kept at the work. When you're obliged to get papers and get your name in the literature – keep your name near the top of the list, you feel that you can't really try crazy experiments. But I was privileged in a way, because at first I didn't have any financial problems and, after 1958, I was in a position in which I didn't really have to publish papers or get

> grants and things. I had a permanent job. And I could do such experiments and I felt obliged to do such experiments. I think that helped me very much in the development of the DNA sequencing procedures.

Fred enjoyed the razzmatazz and excitement associated with the Nobel Award in Stockholm. He invited his family and they were all invited to a ball. 'You get treated more or less as royalty. When you are at home you're just an ordinary scientist. Nobody takes much notice of you,' Fred records.

4

Interview of Fred by the author in 1992: Early life

Fred Sanger (FS) and I (GB) are seated in a recording studio at Imperial College, London. This is a slightly edited transcript of the first interview.

GB *Those of us who have worked with you, Fred, know of your passion and vision for research and your interest in developing new methods. We also know you chose important scientific problems that were well ahead of their time. Less well known, but by no means less important, is your ability to get on with colleagues and students. You guided by example, not by dominating people, always encouraged and rarely criticising. You had, I suspect, a respect for people. You have a disarming natural modesty, rare in one so eminent.*

For this reason many people, including me, would like to know something of your background, your parentage, your schooling and early influences.

FS I was born in 1918 and brought up in Gloucestershire, in a small Gloucester village, Rendcomb, in the Cotswolds, a very nice part of the world to live in. My father was the local doctor; he was a medical doctor. We have a picture of him here (Figure 1, left). Here is a picture of me (Figure 1, right) at about the same time, 1918, being held by my grandfather, Theodore Crewdson, on my mother's side, and another picture of me, 8 months old (Figure 2).

Figure 1. Father, Frederick Sanger (left) and grandfather, Theodore Crewdson with Fred as a baby (right). (Courtesy Peter Sanger and The Biochemical Society, with permission.)

I'll tell you something about my father first. He was, as I said, the local doctor in Rendcomb. He had a fairly wide parish and he had been a scientist at one time. He was medically qualified in Cambridge and after his MB degree he did research for about three years to get an MD. I have actually got a copy of his thesis. This was entitled the 'Biological test for blood considered from the medico-legal aspect'. They knew at that time that blood differed in different animals. They could test for this by the normal precipitin reaction.[1]

GB *He was well ahead of his time, I suspect.*

FS I think so, yes. He worked under Nuttal, who was the first Quick Professor at Cambridge who later founded the Molteno Institute in 1921. But he worked also with Scotland Yard testing blood samples for them and presumably they made quite a lot of use of this particular reaction.

GB *Did he catch any criminals?*

FS I don't know about that. Presumably he contributed.

Figure 2. Fred Sanger, 8 months, April 1919. (Courtesy Peter Sanger and the Biochemical Society, with permission.)

GB *He's the modern Alec Jeffreys.*[2]

FS That's right. Essentially they were very crude tests, but they did not test individuals, of course. I don't think they knew about blood groups then, they were just testing for different species.

GB *But your father didn't continue with research, I understand.*

FS No, that's right. He was a rather religious man and he decided to become a missionary, actually. He had the call. He went as a missionary to China, a medical missionary, so he was attached to a hospital working as a doctor and also spreading the gospel. I think the main thing he did in China was the founding of a school. At that time, only the upper classes – the mandarins, I think, were allowed to be educated. But he founded a school under the auspices of the Church Missionary Society for the lower classes, for just ordinary people. I think in this case he was a pioneer. I think he did regard that as his most important contribution to his life, really.

GB *He sounds a very interesting man.*

FS I think he was. He was a very great influence on me – and my brother.

GB *But he came back from China. Was it to where you were born, to Gloucestershire?*

FS No. He came back because of illness. I think it was probably tuberculosis. He would have wanted to stay on at the school and carry on with the school. But he did come back and I think he first went to Devonshire. His father had died by that time and he set up house there. I don't think he can have been long in Devonshire, but he came to Rendcomb and lived with his mother for a while, until she died in 1913.

Now let me just tell you something about my mother's family. This is my grandfather, Theodore Crewdson (Figure 1, right). We have a lot of information about their predecessors. A lot of them were Quakers, in fact. But he himself had separated from the Quakers. He was a cotton manufacturer.

GB *He looks a jolly sort of person.*

FS Yes, he was. I remember him as a benevolent patriarch, really. His wife had been dead a very long time. In fact, she died when my mother was born. My mother was the youngest daughter. But he was a pretty rich cotton manufacturer living in Styal, Cheshire (Figure 3). How benevolent he was to his workers, I don't know. But, at least he seemed a very nice fellow. He died when I was about 5, I think, and he was the only grandparent I knew at all.

He, however, had bought a property in the Cotswolds. And this was about 8 miles, or even less, from Rendcomb, just across the hills, and near two very small villages. One, called Syde, which was a farm really, and that farm was bought for my uncle, my Uncle Dilworth (Figure 4, standing). He was my favourite uncle, and he farmed that – and there were about 1000 acres, I think. And there was another little village, called Caudle Green, which was just across the valley from Syde – a beautiful little village, and my parents had a little house there, which was a sort of country house where you could go for holidays. That does come into the story a bit later, as we lived there for some time.

Figure 3. Grandfather Crewdson's house, Styal, Cheshire. (Courtesy Peter Sanger and The Biochemical Society, with permission.)

Figure 4. Fred, aged about 3, next to Aunt Daisy, Uncle Dilworth (behind), grandfather Crewdson and brother Theo, about 1922. (Courtesy Peter Sanger.)

Figure 5. Fred's mother, Cicely Crewdson Sanger. (Courtesy Peter Sanger and The Biochemical Society, with permission.)

GB *What about your mother?*

FS Yes, my mother (Figure 5) was the youngest daughter. She had lived the life of a Victorian lady – no suggestion that she would have to work. But they came over to Syde. Actually, they came over in a coach and four, in those days, I believe. My grandfather was rather conservative. There were cars, but I think he came in a coach and four. He kept that going for quite a long time.

GB *Was that fashionable then, even when there were cars?*

FS Well, I don't know if it was fashionable. It was going out of fashion. That's my grandfather (Figure 3) not in a coach and four, but actually he is holding the reins of a trap. I don't think he ever had a car, himself. But my mother, when she was staying at Syde with grandfather, developed a very bad septic finger.

GB *Was that before she had married your father?*

FS Well, yes – and they had to call in the doctor. And, of course, the doctor was my father! Presumably they had to send over a groom to get him. He came over, and that's how it all started! I think they were married in 1916. Both of them were fairly old, relatively. My mother would have been about 36, I think, and my father probably 40 at that time (Figure 6). My brother, Theodore, was born in 1917 and I was born just a little less than a year later in 1918 (Figure 7). I was actually premature – in a bit of a hurry. So we were very close in age and in fact spent a lot of time together. We used to play a lot together. I think I spent more time with him than my parents. One got to rely on him very much. He was a rather different character from me. I was sort of a retiring, quiet, little fellow, and he was much more of an extrovert. He planned the games we played and decided what to do and so forth. It was altogether a very happy time.

GB *Would it be true to say then that you had a good, scientific, medical background from your father, while from your mother's side there was financial security through having a rich grandfather and also a strong Quaker tradition?*

FS Yes, that's right.

GB *But your father had, I understand, also converted to become a very strict Quaker.*

Figure 6. Frederick, father, and Cicely, mother. (Courtesy Peter Sanger.)

Figure 7. Fred (left) and brother Theo (right) as toddlers. (Courtesy Peter Sanger and The Biochemical Society, with permission.)

FS Yes, that's right. By the time I was born he was still Church of England. But the Crewdsons – the Crewdson ancestors – a lot of them had been Quakers. My grandfather had inherited a lot of Quaker books. My father, soon after he had married, started looking through some of these books in my grandfather's collection, and he got interested in it; and it seemed to fit in with his philosophy and he did become a Quaker. All the rest of his life he was very active in the Quakers. We were brought up, essentially, in a Quaker tradition. My mother was not an active Quaker. But she liked singing hymns and that sort of thing, and listening to hymns at least.

GB *It is curious, isn't it, that although the Quaker background is on your mother's side, it was your father who was the actual Quaker influence.*

FS That's right.

GB *How did you think this affected your upbringing? Do you think you had a strict religious upbringing? Did this mean you were treated liberally?*

FS Both really, I think. It was a fairly religious upbringing. We used to have morning prayers and I used to say my prayers with my mother in the evening and this sort of thing. And obviously we did regard that as very important. We never questioned what we were taught and so I think when I came to Cambridge, for example, I was essentially a Quaker; but although I probably could not accept all of the religious doctrines and so forth, I think one was left with a certain morality – moral ideas from the Quaker tradition. One of these things, which I think is important, is Quakers make a very strong thing of truth; telling the truth is very important to Quakers. They believe there is only one truth and you mustn't swear, because taking an oath implies that you don't have to tell the truth normally. And they got into trouble because of that. Quakers would not take an oath in court.

Truth is something that I still regard as important, and I think this is something that is very important to a scientist, because the scientist is essentially searching for truth. It's about truth. I think that I have more respect for truth than most ordinary laymen. For instance, if someone says 'How are you, Fred?' I don't find it all that easy to say 'I am absolutely fine', which is the normal thing to say!

GB *What about your early education. You were taught initially at home, I understand.*

FS Yes, that's right. When I was 5 (Figure 8), we left Rendcomb and moved to Tanworth-in-Arden, another village about 12 miles south of Birmingham. The countryside wasn't so nice, but we were living in the country and we had quite a big garden with a field and a pond. Most of my early memories from when I was 5 until about 17 are from there. I think the main reason for moving to Tanworth-in-Arden (Figures 9 and 10) was that it was nearer to Birmingham which was a very important Quaker centre. My father was again the medical practitioner there. Very early on when we got there he took on a partner. In Rendcomb, of course, he had been by himself.

In the early years we had a governess. She was a Quaker woman and was a very nice person. She taught us, and I think two local boys joined us as well. There were just four of us, at first, at our little schoolroom at home.

Figure 8. Fred, 5 years old, in 1923. (Courtesy Peter Sanger.)

GB *Of course, she wouldn't have taught you any science at that young age.*

FS No, no science. But I think my brother was the one who taught me science, or got me interested in science at that stage, because he was very keen on natural history – animals, looking at bugs, catching newts, etc. We had this pond, which was full of fish, and in the spring masses of toads came in and spawned there. He collected animals. I remember on one occasion we were playing together in a haystack and we suddenly caught sight of a snake just sitting in the hay. I was for beating a hasty retreat, but he just caught hold of the snake behind its head and, in fact, kept it in a cage for a while. It was just a grass snake – a harmless grass snake. I hadn't realised that. He used to keep quite a lot of animals and play around with them and that was, I suppose, our chief interest. He was a great expert at finding birds' nests too.

Figure 9. Mother, Cicely, with Fred and Theodore and baby 'May' (Mary). (Courtesy Peter Sanger and The Biochemical Society, with permission.)

GB *At this young age, would your father and mother have encouraged you in science or would they have been rather distant from you?*

FS Yes, my father, certainly, was interested in science and he would encourage me. I think they did encourage us.

GB *Would there be books on science around the house?*

FS There were books. I remember there were medical books, anatomy.

GB *Fascinating books, anatomy!*

FS Yes, they were. I am not sure whether that was scientific interest or something else. We were rather giggly.

The Downs School, Malvern, 1927–1932

GB *What about the continuation of your schooling after the governess stage?*

Figure 10. Fred (far left) and Theo with mother, Cicely, and young 'May' (Mary) at their house 'Far Leys' in Tanworth-in-Arden, Warwickshire, about 1925. (Courtesy Peter Sanger and The Biochemical Society, with permission.)

FS That was fairly traumatic, really. I was sent away to school, at the age of 9, to a prep school to a boarding school. That was quite a break. I was pretty homesick at first and didn't enjoy it very much. But, you know, you settle down and you accepted things. It was at the Downs School, near Malvern. It was a Quaker school, which was presumably one of the reasons for choosing it, and had a very enlightened headmaster, but little boys are little boys. There was a good deal of bullying, I suppose, and one was not terribly happy. I don't think I was bullied any more than others.

When I was at this school I was called 'Mouse' and I think this picture probably illustrates that I looked like a little mouse (Figure 11). That's little May (Figure 11, centre), my younger sister, who was born five years after me, so there is a much bigger gap between her and me.

GB *I caught a glimpse of your father here, and your resemblance to your father in one of the pictures.*

Figure 11. Fred (left), sister May (centre) and elder brother, Theo (right). (Courtesy Peter Sanger and The Biochemical Society, with permission.)

FS That is quite a nice picture (Figure 12) of us in the pannier. That was our method of transport. Instead of going in the pram we went in the pannier side by side.

To return to the school, I think academically I was sort of average. I was not one of those people who was always top of the class, but I was near the top. I was fairly bright.

GB *Have you kept any of your school reports, or did your father ever give you your school reports?*

FS We used to look at them; they were quite good, really. I was quite a good boy! You see my brother and I started being educated with the governess at the same time. So we were really at the same level. But he, of course, went to school a year before me. He tended to be towards the bottom of the class while I tended to be near the top.

GB *You found work quite easy?*

FS I quite enjoyed it, especially the more mathematical work. I didn't enjoy Latin much. I never did Greek.

Figure 12. Theo and Fred (behind) with father, Frederick, and Jimmy, the donkey. (Courtesy Peter Sanger and The Biochemical Society, with permission.)

Bryanston School, 1932–1936

GB *From 1927 to 1932, up to age 14, you were at this Malvern prep school. Then, where did you go?*

FS I went to Bryanston. It was a very new public school. It had only been going for about four or five years when I went there in 1932. And I enjoyed it much more than the Downs School. It was much more liberal and there was a more friendly atmosphere with the elder boys. And that was enjoyable. Academically I did fairly well there, actually. It had a different sort of system of teaching regime. It was a new system – the Dalton system, where you didn't spend so much time in classes and you had to work by yourself. You had classes but they were not quite so frequent and you had assignments, which was like prep, where you went off and did things by yourself. You did this in ordinary school hours. So this left you rather free to choose what you did. This really suited me, because I was prepared to work, and really enjoyed the work, so I did work hard. On the other hand people who didn't want to work did not do particularly well there.

GB *Presumably by this time you must have been taught biology and chemistry?*

FS Yes, that was my first introduction to real science.

GB *Were the teachers good?*

FS I think they were good, and I certainly enjoyed, and did pretty well in, scientific subjects. There was one particular teacher, Frazer Hoyland. He was the brother of the headmaster of the Downs School, where I had been previously, and he was the biology master. Coming from the Downs I think we had a special relationship with him. He used to take us for expeditions in his car for example. They had, I think, what was called a 'Biological Society', which was like having hobbies in free time. I belonged to that society, and we used to mess about in the lab making slides mainly, and looking at them under the microscope.

GB *It sounds as though it suited you well. What about chemistry?*

FS Yes, that was taught by another master, who was in fact my housemaster, a Mr Ordish. He was a rather quiet, unassuming, slightly eccentric sort of person, but I got on very well with him. Now, when I was in my final year there, in 1936, that was quite a year for me. I had already done the School Certificate – that's the first exam you took, equivalent to 'O' level.[3] I managed to do well in that. I got seven credits. I was quite exceptional, and was called 'the seven-credit Fred' for a bit, but it meant I could get straight into Cambridge. I had the qualifications to get there – and that was the next hurdle to get over. My father had been at St John's College, and he wanted me to go there. I got officially entered for that.

GB *He was not the only member of your family who had been to Cambridge, I understand.*

FS Yes, my Uncle Tom went to Cambridge. His real name was Hubert. He was quite a lad at Cambridge. He was captain of the boats, and was captain of the Cambridge Eights, I think. He was a great rower. He had been fairly famous and everybody remembered him, whereas my father was just an ordinary MD.

GB *So you didn't have to gain a scholarship or exhibition, or even pass an exam to get into Cambridge?*

FS No, at one stage I sort of toyed with the idea of trying to read for a scholarship, but I went and discussed this with the physics master – I got rather ambitious, and he said 'Well, you know, Fred, you're not quite up to it!'

So in my final year, in 1936, I had a free year and I could spend quite a lot of time at science, and Mr Ordish – the chemistry master, he had done research in his time, and he used to mess about in the lab and do a few little projects – mostly with dyestuffs. He allowed me to come to the lab and give him a hand with these projects. So I went to the lab and spent a period in the lab making these beautiful, coloured crystals. I enjoyed that very much. Working in the lab was a marvellous change for me from just sitting and studying books. I decided that this was something I really enjoyed.

I should say, at this time, I had become rather more independent of my brother. My brother was more interested in animals and had, in fact, a very fine collection of skulls, which was his passion at this time. He used to collect these animals' heads and boil them up to get the flesh off them – making terrible smells – my mother finally disapproved of this. I became more interested in making things with my hands – in particular carpentry – we learned a bit of carpentry at school. I had a workbench at home with some tools and I used to make things in the holidays (Figure 13) while my brother would go off and collect animals' heads. This has been an interest of mine, really, doing things with my hands. I also had a forge – a blacksmith's forge (Figure 14) and I used to do some ironwork too in the holidays.

GB *How did you get the necessary temperatures to heat up the metal?*

FS You could bend the metal and weld it with the forge and the anvil. I'm with my sister (Figure 14) – she's blowing up the bellows! I was about 15 or so then. The biggest thing I made was the gate, and that was in our garden for quite a while.

GB *Are you related to Ruth Sanger FRS,*[4] *well known for her work on blood groups?*

FS Yes, she's my first cousin (Figure 15). She is the daughter of my Uncle Tom, who emigrated to Australia. She was born in

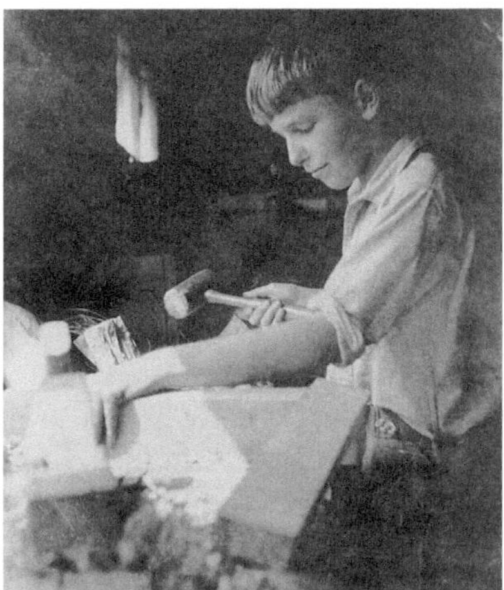

Figure 13. Fred, aged 13, at home at his workbench. (Courtesy Peter Sanger and The Biochemical Society, with permission.)

Figure 14. Fred at his forge with May, 1933. (Courtesy Peter Sanger and The Biochemical Society, with permission.)

Figure 15. Ruth Sanger FRS (1918–2001) in 1972. (Copyright Godfrey Argent Studio.)

Australia. Then she worked with Rob Race on blood groups and finally married him. So there is some science – some DNA, in the family.

School exchange visit to Salem, Germany, 1936

GB *Fred, just before going to university, I understand you were on an exchange school visit.*

FS That's right. I had this year more or less to spare and I took a half a term exchange visit to Germany – that was to a German school, Salem, which is a school that was founded by Hahn, the same chap who later founded Gordonstoun. It's the same sort of system – a very modern sort of school. They had a rather close connection with Bryanston. That was quite an experience. I had learned German – a certain amount, in school. It was one of the subjects I took for School Certificate but it was a bit traumatic at first. I couldn't really understand spoken German and I couldn't carry out a conversation. But I gradually settled down and it was an interesting experience, because this was the height of the Hitler period. Of course Hahn had been chucked out because he was a Jew and Headmaster and there was a lot of Hitlerism going on there, and that was interesting. It was an interesting school in that the general system was very different. There was a great accent on sport and leisure activities and we got the chance to go out on expeditions from the school. There was another boy from Bryanston, David Forbes and myself, the two of us. So we spent quite a bit of time together. We weren't meant to talk to one another in English and tried to talk German together. We got to know quite lot of German boys.

GB *Were you made to take part in the standard things that the Nazi youth were meant to do in those days?*

FS Some we were exempted from, and some we didn't take part in. It was usual to stand up and 'Heil Hitler' before every class. We stood up, but didn't 'Heil Hitler'. We were allowed to be excused from that. The people were quite reasonable – most of them. There were one or two people, who were rather the Prussian types, and some of the masters, who used in their days off to dress up in their black shirts and their brown shirts and go off and do the sort of activities they did.

GB *Were you political enough to be frightened by this sort of thing, or have some sense of warning. Were you a political animal?*

FS I wasn't very political. You know, I felt, we had to have collaboration between different nations. I suppose I still had the Quaker philosophy that all men were brothers. People were very nice and the people I got to know – some of the boys were particularly friendly and weren't all that keen on the Hitler system. They had 'Hitlerjugend' exercises and things to do.

GB *Did you, yourself, see any signs of impending war? Did you hope this was a phase that would disappear?*

FS Yes, I think so. I don't really know. There was terrific enthusiasm, terrific nationalism. One thing struck me, rather. We used to have prayers every morning – a reading by the headmaster – and the headmaster was quite a nice sort of person, a quiet fellow, rather intellectual, and friendly. Every morning he would read a piece, either from the bible or from *Mein Kampf* and this rather sort of stuck in my gullet, being a Quaker.

GB *It sounds to me you were detecting that something was wrong!*

FS But you did not think about it, too much. But it was quite an interesting experience. David and I used to go out at weekends on bicycles, usually borrowed bicycles, and explore the countryside. We saw a lot of south Germany and were taken 'prisoner' at one stage, which was a bit alarming! That was with two German boys. We went off camping. We camped in the grounds of Spetzgart – a school that was the equivalent of Salem. It was actually a mixed school with not many girls, and we camped in the grounds. When we woke up in the morning we found ourselves surrounded by these hefty girls, and this was rather alarming. But we managed to escape into a nearby wood. Unfortunately they had got our bikes and our camping gear and everything else and we were at a bit of a loss as to what to do. We decided to creep up and get our bikes back. We eventually saw our bikes and made a dash for them, but just as I was starting off, I was captured by the janitor, I think. I was put in 'prison' – the place where they kept their shoes. It turned out there was some sort of feud between the two schools, between Salem and Spetzgart. Germans love this sort of thing – mock battles! I was locked up in this room with the shoes of the girls, and I took what they called the shoe spanner – something you put inside a shoe to stretch it – and I managed to use this as a screwdriver to get the lock off the back of the door. Luckily, I just got out of the door as the guard, who was guarding me, was out of sight. I dashed off into the woods and managed to hitch-hike back to Salem. Apparently, I became quite a hero at this stage, and was hailed as 'Held der Tages' – The Hero of the Day. I think I had more or less been held as a hostage, you see. That was quite an excitement for me!

Cambridge University undergraduate, 1936–1940

GB *Shall we move on to your undergraduate time at Cambridge? What were your first impressions of Cambridge? Who was your first tutor – at St John's, you say?*

Figure 16. Fred, at Cambridge, aged 18. (Courtesy Peter Sanger.)

FS Yes, I was at St John's. My moral tutor was actually Robert Howland.[5] When I came to Cambridge (Figure 16) I had to decide what subjects to take. Originally, while I was still at school, I had intended to study medicine. But about a year before I went to Cambridge I decided, well, I did not really want to study medicine – I had seen what it was like. My father had a very busy life. He had to rush around from one person to another, and could not really concentrate on anything. I didn't think this was quite the life for me, so I decided I would study science. So when I got to Cambridge, I took physics and chemistry – that was a rather obvious thing. I looked around for another subject. You had to take three subjects for the Tripos.[6] There was a subject called biochemistry. This was something I had never heard of before. This was a half-subject and I had to take another half one as well. So I went to see the supervisor in charge of biochemistry at St John's – that was Ernest Baldwin – and he was a very enthusiastic man (Figure 17, left). He told me all about it and I thought this was terrific. Fancy being able to explain biological

Figure 17. Left: Ernest Baldwin (1909–1969) (The Biochemical Society, with permission). Right: Frederick Gowland Hopkins (1861–1947), Royal Society portrait (by kind permission of the executors of the Meredith Frampton estate).

systems in terms of the exact science of chemistry. I had done quite a bit of chemistry – organic chemistry, too. And I decided, yes, I would do that. And that was how I first came to be in contact with biochemistry, and that was the subject I did best in, all the way through, really.

GB *This was in the Part I. What other subjects did you do in the Part I?*

FS Well, I took maths, actually, as a half-subject. It was a bit of a struggle, but I managed it. I'll tell you what happened, after the first year I didn't do very well in physics or in maths, actually, and I found that they were a bit beyond me. The thing was that most of the students at Cambridge had done the Higher Certificate at school. That meant they had done two years' solid work after the ordinary School Certificate, and in a subject like physics they were well ahead of me, and I just couldn't catch up. So I decided to switch over to physiology instead of physics. So then, from the second year on I decided to do physiology, chemistry, biochemistry and maths. And I found physiology pretty easy. I think the higher maths was a little beyond me. Since then I have always been scared of mathematical formulae.

GB *But Fred, you obviously did well enough because you went on, I understand, to do the Part II in biochemistry.*

FS Yes, but after three years. You see it took me three years to do the Part I Tripos. Normally, you should have finished by then and have got a BA. I got the BA after three years, but I decided I would like to try and do a Part II in biochemistry – just nothing else but biochemistry, for a fourth year.[7]

GB *What was the course like then?*

FS It was very good, and very exciting then; it was 1939, right at the beginning of the war. Yes, this was in the Department of Biochemistry in Cambridge – which was a relatively new department founded by Gowland Hopkins (Figure 17, right), and it was full of his young disciples, you might say, all rather keen people in a new subject making new discoveries largely vitamins, enzymes and this sort of thing. No genetics or anything like that. But it was an exciting place to work and in the Part II class you came into contact with all these young, enthusiastic people.

Figure 18. Malcolm Dixon (1899–1985) aged about 50. (Copyright Ramsey and Muspratt, with permission.)

I think it was a little, you know, too concentrated. They all wanted to talk about their own work and their own subject. Their lectures were very high-powered.

GB *Did you have lectures from Dixon?*

FS Malcolm Dixon (Figure 18), yes, very fine lectures. I mean, rather dull, but well organized, his lectures were. We had Joseph Needham (Figure 19) on morphogenesis – that was very exciting. He again went much too fast – terrific lecturer, he could rattle away in several languages at once! Who were the other lectures, yes, Bell was another good lecturer. But I think the best lecturer was Ernest Baldwin, but he was more a Part I lecturer. He was very good for the Part I. A very inspiring person, I had quite close contacts with him. He was interested in comparative biochemistry, that was his special subject.

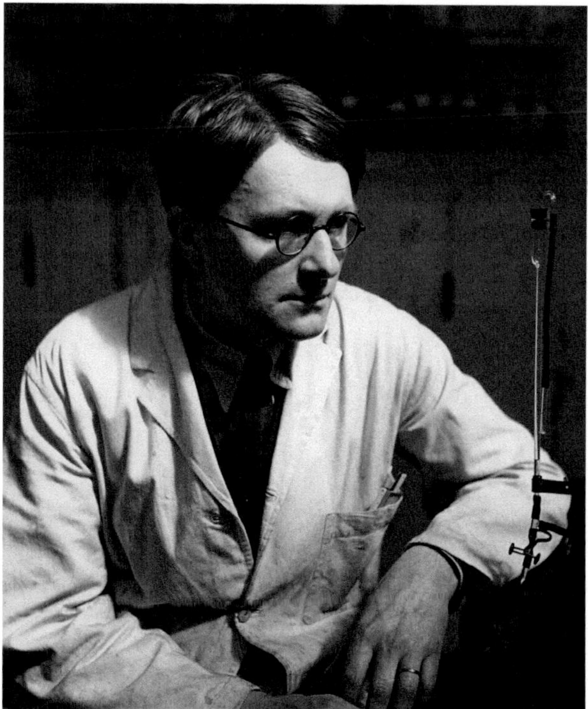

Figure 19. Joseph Needham (1900–95) in the Biochemistry Department, Cambridge, 1937. (Courtesy Stephen Burch.)

GB *Of course, a young man in Cambridge has more to do than simply work! This was also the beginning of the war; what were your impression of Cambridge? What were your other interests at the time?*

FS Well, that's true, yes. I did work hard and well, an important interest was this Quakerism, really. I was still pretty keen on Quakerism, myself, and a lot of my friends were Quakers. And of course, Quakers are pacifists, they won't fight, and I was a keen pacifist. Of course, this wasn't quite the time to be a pacifist. I was a conscientious objector and I got unconditional exemption from military service, and so I was lucky in that respect. But I also had joined various pacifist societies, particularly the Peace Pledge Union,[8] which was founded by Dick Sheppard – and I had signed the pledge that I wouldn't fight, so I felt obliged not to fight, and at that time I felt it would be entirely wrong going around killing people.

Figure 20. Joan Howe, 1940. (Courtesy Peter Sanger and The Biochemical Society, with permission.)

GB *Was your position accepted by officialdom?*

FS You had a little trial, as it were, and a committee asking you questions and they decided that I was genuine.

GB *You had some sadness to cope with during your undergraduate years – both your mother and father died, then.*

FS Yes, that must have been in my first year, I think.

GB *It must have been quite a shock for you as a young man when both your parents died within such a short time.*

FS Well, it was really. My father and mother both had cancer. My mother actually had it first, but my father got it, and he had it rather badly and decided to have an operation. He died in the operation, rather suddenly and I hadn't expected it, so it was quite a shock. Then my mother died a bit after that – rather more slowly, which again was rather trying.

GB *Did you meet your future wife about this time?*

FS Yes I did, and that rather eased the shock of losing my parents. Perhaps I should tell you about that. I joined what was called the scientists' Anti-War Group – scientists who were against the war at this time. This must have been just before the war, and the

Figure 21. Marriage to Joan Howe, December 1940. (Courtesy Peter Sanger and The Biochemical Society, with permission.)

group decided we should write a report on the economic effects of rearmament. I was chosen to organise this report. Really, I hadn't a clue what to do but anyhow I made an effort and we decided to get people together to do this. We decided we needed an economist and among us were some Newnham College students, who were part of this anti-war movement. One of them said, well I have a friend, Joan Howe, who is an economist, and she brought her along to this meeting (Figure 20). That was how I met Joan Howe, who finally became my wife (Figures 21 and 22). I got to know her pretty quickly, and we spent quite a lot of our spare time together. We were courting at this rather critical stage when I was doing the Part II Biochemistry.

GB *You managed both!*

FS Yes! (Figure 23)

Figure 22. Wedding of Fred and Joan Howe, December 1940. (Courtesy Peter Sanger.)

Figure 23. Fred and Joan, about 1942. (Courtesy Peter Sanger and The Biochemical Society, with permission.)

5

Interview of Fred by the author in 1992: Insulin and the Biochemistry Department, University of Cambridge

Fred Sanger (FS) and I (GB) are seated in a recording studio at Imperial College, London. This is a slightly edited transcript of the continuing interview.

Postgraduate research at Cambridge, 1940–1943

GB *I've read, Fred, that at this stage you had still not quite decided to go into research.*

FS Well, the thing is, I had not really decided that I was that good. I hadn't been getting first classes in my Part I exams, and so I really didn't have the confidence that I could do research. Usually you had to get a first to go into research. In this Part II Biochemistry there were all these brainy people who had got firsts in their Part I. They all seemed very clever compared to me. So at the end of the year I took the exams and I sort of went off and didn't think too much about what I was going to do. In fact, I was thinking I would probably do some war work of some sort. War had broken out. But then, to my surprise, someone had seen in the paper that I had got a first. Actually, I was staying with my cousin. He was in the Air Force and he had been to the mess and he had just had a look at the Cambridge results in *The Times* and he'd seen I had got a first. And I said that's ridiculous, it couldn't be, but it was so. I had got a first. There were just two of us who had got firsts in this exam. So that changed my mind. So there was the possibility of doing research.

GB *But before you actually started research you had actually preplanned some training in Devon, I understand.*

FS Yes, that's right, I went to a place called Spison (now Spiceland), a Quaker Relief Training Centre, which was a training place for conscientious objectors, where I was learning various things they could do to help save lives, instead of taking lives. We learned agriculture and a bit about building, and after I had done that course I went to work in a hospital, cleaning the floors and that sort of thing – just as an orderly. I started work there. That was the first job I had. I got 10 shillings (50p) a week for that.

GB *Was that enough to live on?*

FS Well, I suppose it was. We had our accommodation taken care of in the hospital. There's a picture of me cleaning the floor in a toilet (Figure 24). That's me working hard!

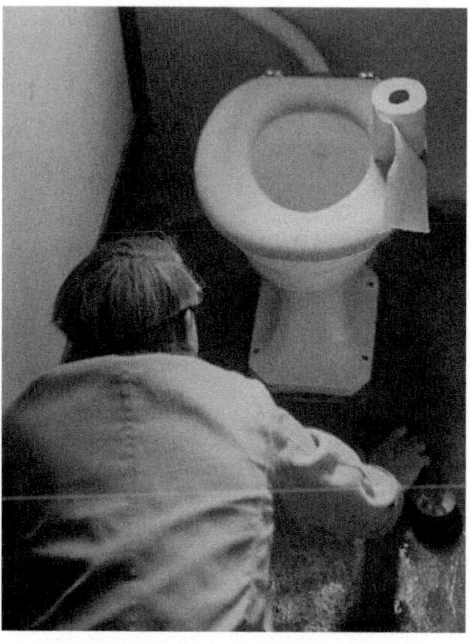

Figure 24. Fred at Winsford Hospital, near Bristol, cleaning the toilet, 1940. (Courtesy Peter Sanger and The Biochemical Society, with permission.)

Figure 25. The 'old' (left) and 'new' (right) Department of Biochemistry, University of Cambridge. Both are currently in use.

GB *I don't think anyone would recognise you from that back view.*

FS That was not very inspiring work. But that was about the time of Dunkirk. When the people were coming back from there they had very nasty wounds. It brought me in contact with real life. But I had decided I didn't really want to spend my life doing this and I thought I would try and see if I couldn't get back to Cambridge and back to the lab. I wrote to Professor Hopkins, Gowland Hopkins, who was the head of the Biochemistry lab (Figure 25), and asked him if it would be possible to come back and do research. I was in a favourable position in that I was a conscientious objector and was free to do that. And I had money – I didn't need any pay because my mother had been fairly rich and she had left me enough to keep me going. But I didn't get any answer from Hopkins at all, or anybody. So after a couple of unanswered letters I decided to go to Cambridge, and take a bit of leave. I found there were quite few people there who would be pleased to have me working with them. It was a little bit difficult to decide between PhD supervisors. I think the two main contestants were Ernest Gale – he had been my supervisor at one time – and Bill Pirie,[1] who was a protein chemist – rather a wild boy (Figure 26, left). He was rather persuasive and he convinced me I should work with him. I thought that did sound rather more exciting than Gale's work which seemed a bit methodical and not that inspiring.

GB *So you started in October 1940. The war was on. Was it a difficult time?*

FS Yes, it was, but in a way I was protected. We never had very serious bombing in Cambridge. There was one bomb, I think, and blackouts, food rationing and all that sort of thing. But, Bill Pirie, he was interested in trying to make edible protein from grass. And he remained interested in that – it was his main thing. He started off by giving me a bucket of frozen, mashed up grass – a huge bucket of this green stuff. He believed, if you had a PhD student, in putting them in at the deep end, and he said, 'You want to work on this.'

GB *At least he did not make you mow the lawn and get your own grass. Was it a special grass, or any old grass?*

FS As far as I know, any old grass. But I didn't know what to do – it took about a week for this 'stuff' to melt before I could get working on it. And then, after about a month, he left and went on to another job at Rothamsted Experimental Station! I never really did anything with him.

GB *And so, who took over as your PhD supervisor?*

FS I was very lucky, it was Albert Neuberger (Figure 26, right). He was a very knowledgeable person. He gave me a project and I worked with him for three years. This was work mainly on lysine metabolism. The main project I was working on was trying to

Figure 26. Fred's PhD supervisors: N. W. Pirie (1907–97) (left) (copyright Godfrey Argent Studio) and Albert Neuberger (1908–96) (right) (The Biochemical Society, with permission).

synthesise the keto acid[2] corresponding to lysine, which was supposed to be the first degradation product of lysine. I did a lot of straight chemistry which was what I enjoyed. I got a lot of help from Neuberger – I was the only PhD student he had at the time – we worked more or less together. I regard Albert as my main teacher, really. He taught me how to do research and a lot about amino acid chemistry. He was very knowledgeable. He worked not only on amino acid metabolism but he had worked on glycoproteins[3] and on porphyrins[4] as well. He was a very kind person too and very helpful to me. I feel I owe a lot to him.

GB *Fred, you have told me your thesis had the title 'The metabolism of the amino acid lysine in the animal body'. Did you work hard? Were you interested in the work?*

FS Oh yes, I worked pretty hard. I got involved. When I am on a project, it seems to be the main interest of one's life and you get stuck into it. Particularly for the keto acid, that was the thing that interested me most – I don't know why – it never worked. I never managed to make the keto acid. We did other things on the side, and in fact we got four papers out of it. One thing we did, as well as the metabolism of lysine, was essentially war work, analysing nitrogen in potatoes, and what sort of form the nitrogen was in.

GB *But you wouldn't have been required to do this, you would have been exempted?*

FS And I think Neuberger was also exempted. But it was really Neuberger's decision more than mine. At that stage you more or less do what your supervisor suggests. Unlike you, of course!

GB *We'll come to that later!*

FS But I did, certainly. I had never done any research before and I followed – more or less – Neuberger's suggestions.

GB *And what about your thesis itself, was it a pretty chemical thesis? Did you learn a lot about the chemistry of lysine?*

FS Yes, I did, particularly lysine, as you say, and this did of course come in useful later particularly in the work on protein sequencing. I knew quite a lot of chemistry and I had quite a lot of experience.

GB *And why did you choose lysine?*

FS I didn't really. I think Neuberger chose it. But I think it was an important amino acid and it was an essential amino acid. It was

Figure 27. Fred's PhD examiners: Charles Harington (1897–1972) (left) (copyright Godfrey Argent Studio) and Albert Charles Chibnall (1894–1988) (right) (copyright Lafayette Photography).

one of the ones that did seem to be limiting in incomplete diets – so it was quite an important one, and the course of its degradation and metabolism was not known. So Neuberger was interested in this.

GB *When did you get your PhD?*

FS In 1943. I wouldn't say it was a great thesis. In fact, I was examined by Charles Harington (Figure 27, left) and Chibnall, Professor Chibnall (Figure 27, right). Harington was rather critical of it, but it was mainly a question of spelling mistakes because I had typed it all myself and I am not that good at spelling. But they did decide that if I corrected these mistakes in the thesis, I could have a PhD.

GB *You've spoken about some influences of the war in this period. Did you have to fire-watch?*

FS We fire-watched, that was a contribution to the war effort. We spent a night a week. I don't think we had to do it. It was voluntary but I was willing to do it. Usually two of us spent the night sleeping in the lab. This was quite interesting. You met various people. I remember particularly fire-watching with Joe Needham, which was the closest I ever got to him. He was a grand man. It was quite nice just talking to him (Figure 19). He was a man of very wide knowledge and very liberal ideas.

GB *Had he by that time become interested in China,[5] or was he still working on strictly biochemical problems?*

FS He was working in the lab, yes. He talked about China, a bit. I think he was gradually beginning to move over. I found him a very impressive character. He was Reader in the department, second to Gowland Hopkins, but he didn't spend time doing any organisation. He just gave lectures in the Part II.

GB *And what were your general impressions of the Department of Biochemistry, thinking more widely than your immediate supervisor? Was it a friendly place? Were there undercurrents, or what?*

FS They were very friendly, quite open about their work, and always very enthusiastic and that was a nice thing. Hopkins himself was of course a very great influence (Figure 17, right). He was very old at that time, rather frail, but he used to totter around the lab and he used to talk to me. Although he spent much of the time grumbling that his eyes were not very good, and that he couldn't hear very well, it was in fact inspiring.

GB *Fred, when you met Hopkins, when he came round the lab from time to time, he spoke to you, I understand, about the essay you wrote in the Part II Biochemistry exam?*

FS Yes, that was quite an occasion, really. In the exam at the end of the Part II one of the questions was an essay question – you were given a number of titles – and you had to write for three hours on this subject. One of the titles was 'Comparative biochemistry'. I was quite good at this, so I wrote a standard essay, but I found I had got quite a lot of time at the end of the essay. So I started to develop the idea, instead of the idea of the survival of the species, it was more the idea of the survival of the chemicals – the survival of the fittest molecules. I think this idea was at that time rather a new one. I wrote rather extensively on this. I think I had thought of this idea during the exam. I remember Hopkins, when I met him again when I came back to Cambridge, said that he had been particularly impressed by my essay, and of course this rather pleased me. I think this did help me considerably in getting a place in the lab.

GB *Did this influence you, when you became head of lab, years later? Did it give you clues as to how to run a lab successfully?*

FS Well I suppose it did, I don't know. Subconsciously I don't think I ever thought about running a lab. I was always just interested in the research. I suppose one got one's ideas from the other people in the lab.

Early studies on insulin, composition and amino acid sequences

GB *Turning to the next and the first major period of your research – your work on insulin, I believe you started this in 1943. Could you give us some idea of how you became interested in insulin, because this in fact was a break from your thesis work on lysine, wasn't it?*

FS Well, yes, this was a question of a new job, really. Neuberger left the lab for the National Institute for Medical Research in London, and I had finished my PhD, so where would I go next? Well, at this time Chibnall was appointed as the new Professor to succeed Hopkins. Hopkins, although he was officially head of the lab, had not been very active during the previous years. He was very old, and the lab was sort of run by various people, with various degrees of competence. In fact, Chibnall, when he took over, found the lab was in a pretty bad state. The main administrator, the main secretary, had to be sent to prison! He had been taking a little bit off the lab funds. So it was a bit chaotic and 'Chibs' had a bit of a problem straightening that out. But the lab went on alright in spite of all this neglect. It was driven by the enthusiasm of the science, and the scientists.

Chibnall was an expert on amino acid analysis and was interested in protein structure. He came with a group of people who worked with him to Cambridge (Figure 28), and I was looking for a job. He offered me a job with him which, I think, at first was a job paid by the Medical Research Council. This was the first time I had any money because during my PhD I did not have any pay at all. I just lived on the money I had inherited from my parents.

Figure 28. Department of Biochemistry, Cambridge, 1950; Chibnall front centre; on his left Malcolm Dixon (1899–1985), Ernest Gale (1914–2005) and Ernest Baldwin (1909–69); on his right Joseph Needham (1900–95). Fred Sanger, fourth row, extreme right. (Courtesy Peter Sanger and The Biochemical Society, with permission.)

GB *Did Chibnall suggest the insulin problem, or did you actively seek that problem yourself?*

FS No, I didn't really pick that problem by myself. I think I was drawn into it by him. Chibnall had been working – he never did work in the lab himself, he was an organiser – but his team had done work on insulin. They had chosen insulin because obviously it was an important protein and it was one you could obtain in a pure form. You could get it from Boots (a pharmacy) in a bottle. They had done a very accurate analysis of 'free' amino groups,[6] which was the main way to analyse proteins then. They found there were more free amino groups – many more – than could be accounted for by the ε-amino groups of the lysine residues. This meant that insulin had free α-amino groups at the ends of rather short chains.

GB *And who suggested your first problem – a method to detect the N-terminal amino acids of insulin?*

FS Chibnall did. He had done this, you see, and he had found these amino groups in insulin. He suggested I should identify them – obviously work out a method for identifying them. I think if I had had a choice of what I wanted to do at that stage I would have gone on working with lysine metabolism. Once you get into a project you get interested in it. Of course, that would have been a mistake, I think. I would have probably never have done anything very significant.

GB *You might have discovered the Krebs Cycle! Anyway, you studied the free amino groups of insulin, but what was the first reagent you used – it wasn't the one that you're famous for, was it?*

FS No, I played around with different reagents. The first one I tried, for quite a long time, was the reaction with methanesulphonyl chloride forming methane sulphonyl derivatives of amino acids.[7] The reason for this, I think, was because I wanted to get something that I could fractionate by partition chromatography. I think the most important discovery in this field had been the discovery of partition chromatography by Martin and Synge (Figure 29). They had discovered this method,[8] which was a method of fractionation

Figure 29. Archer John Porter Martin (1910–2002) (left) and R. L. M. Synge (1914–94) (right), co-inventors of partition chromatography. (Copyright Godfrey Argent Studio.)

far above anything that has previously been discovered, particularly for water-soluble substances. They had discovered this method for separating the acetyl amino acids. This made it possible to analyse, reasonably accurately, the mono-amino acids, alanine, valine and leucine. Tristram, who was one of Chibnall's 'gang', had got this method developed, and was analysing proteins using partition chromatography. I could see that this was a great advance. It made possible chemical work and fractionation of protein derivatives and peptides.

GB *So you made your derivatives, acid hydrolysed them and fractionated on the Martin and Synge partition chromatography in chloroform.*

FS That's right, mainly in chloroform, with the dinitrophenyl stuff. With the methane sulphonyls, I thought they would be very similar to the acetyls and so that's why I really chose them in the first place. But there had been work with the chlorodinitrobenzene labelling of end-groups and free amino groups, and Chibnall had told me about this. It was by Abderhalden.[9] They had shown that it was possible to label with chlorodinitrobenzene, but you needed to heat to get the label – the dinitrophenyl (DNP) group – to go onto the protein. That would not be any good for end-grouping because it would hydrolyse a certain amount of protein. But I made a lot of these DNP derivatives with chlorodinitrobenzene and I found that they could be fractionated well on the columns, which was necessary for any identification.

GB *So you describe this in your first paper on insulin published in the* Biochemical Journal *in 1945.*[10]

FS Yes, the point is, however, that chlorodinitrobenzene was not reactive enough, so we had to make some other derivative. There were two possibilities; one was trinitrobenzene which I actually made and it worked fairly well. The other possibility was fluorodinitrobenzene, which was much more reactive – it would react at room temperature. We didn't have any, but Chibnall discovered (he wrote a little article about this in the *Biochemist* later), while having a glass of beer with a chap in the chemistry lab, that D. C. Saunders had been making fluorodinitrobenzene as part of his contribution to the war effort. I think he made many fluoro compounds, I don't know whether as poisonous gases, or maybe as

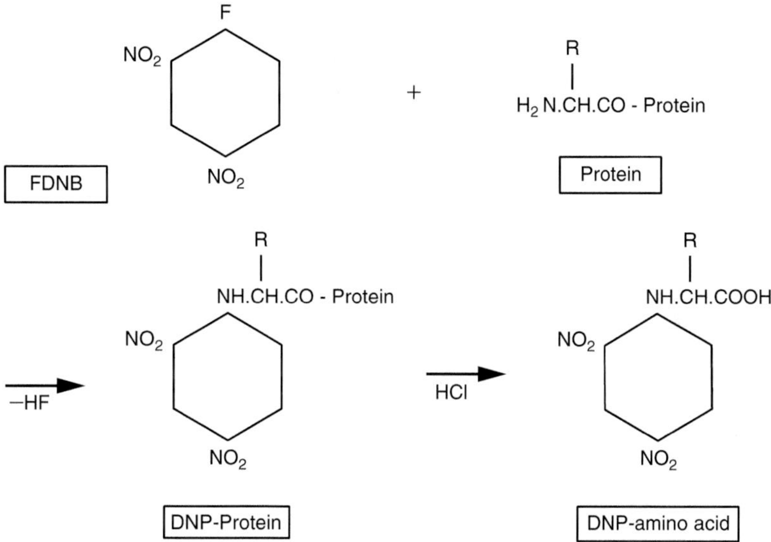

Figure 30. Chemistry of the Sanger method: the fluorodinitrobenzene (FDNB) method of labelling the N-terminal amino groups of proteins giving rise to dinitrophenyl (DNP) protein. Subsequent hydrolysis with hydrochloric acid (HCl) releases a dinitrophenyl (DNP) amino acid. (The Biochemical Society, with permission, modified.)

antidotes to poison gases. Anyhow, he had a good supply of fluorodinitrobenzene, and we got some of this and that did the trick (Figure 30). We could label up the insulin, make the DNP insulin, under mild conditions and then hydrolyse with acid and obtain the DNP-amino acids. On insulin we found there was DNP-glycine and DNP-phenyalanine. So we had identified the two end-groups and that accounted for all the α-amino groups.

GB *At that time, I understand the molecular weight of insulin was not known, or indeed it was believed to be higher than it eventually turned out to be.*

FS That's right, yes, if you measured it in the ultracentrifuge, which was the great thing in those days, you got a value of 48 000. If you measured it by X-ray crystallography – the size of the unit cell – it was 36 000. So quite a big protein, and very much larger than we thought it was. But obviously the results didn't tally, so the assumption was that it was 12 000 and these were different aggregates; 12 000 would fit both results. So quantitatively, there were two phenylalanines and two glycines in this 12 000. This was what we thought at that time. This was based on Gutfreund,[11] who was using osmotic pressures to measure molecular weights, and we sort of followed that line assuming there were four chains in the molecule.

GB *So this would have naturally led you on to thinking about separating these chains.*

FS That was the next idea, yes. That was the next thing we succeeded in doing and these chains would be joined together by disulphide bridges of the cysteine residues. That was an assumption but it seemed the obvious thing, because this was the only amino acid that could be thought of as making a covalent bond between chains.

GB *And how did you succeed in separating the chains?*

FS First of all we had to break these disulphide bridges. We did that with performic acid converting the S—S bonds linking the cysteines to sulphonic acid groups and this gave us nice stable products, which then had extra acidic groups in them. We tried various methods for fractionating the chains but we found that the easiest things were classical precipitation methods with various solvents – non-chromatographic methods. We did play around with chromatographic methods. It was fairly easy to get what we called a fraction A, which was very acidic and just had the glycine end-groups, so we thought that was one of the chains. The yield of these chains was not very high, so we couldn't be sure that all the glycines were there.

GB *But you had plenty of starting material.*

FS Well, by present-day standards, we got a gram at a time, so we had to work with milligrams of the isolated chains, which was a small scale for those days. We didn't use 'bucket chemistry'. Of course, the partition chromatography methods were small-scale methods, and that was one of the great advantages of them. But we did manage to get two fractions out, fraction A, which had the glycyl end-group and fraction B, which had the phenylalanyl end-group. That was successful.

GB *Then you – the next stage in your career – obviously took the decision to try and sequence.*

FS Yes, well, (hesitantly) again this was an extension of the DNP method. We knew that if you did a complete acid hydrolysis you got DNP-amino acids. But if you would do a partial hydrolysis you could get peptides, and this looked like some way of extending your knowledge of the sequences in insulin. In fact, this was a way of getting the first-ever sequence of the protein. This was fairly straightforward. Instead of using strong acid for a long time, we used weaker acid and we were able to isolate various peptides. Here is a typical column. It is a silica gel partition column (Figure 31). You took the various coloured bands and you hydrolysed them with acid and you got DNP-amino acids and

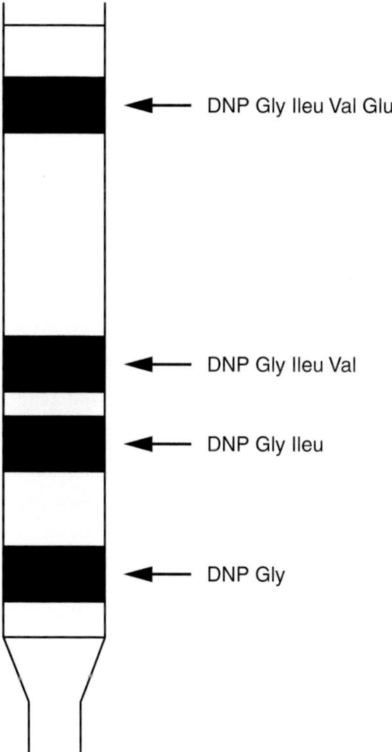

Figure 31. Silica gel column partition chromatography of a partial acid hydrolysis of DNP-insulin A chain. This shows the separation of the yellow-coloured DNP glycine and three yellow DNP peptides. (The Biochemical Society with permission, modified.)

other amino acids. These amino acids were detected by paper chromatography. Paper chromatography had just been invented at that time and this was another great advance.[12] It was an offshoot of partition chromatography. You could just spread things out on a two-dimensional system, spray them up with ninhydrin (a reagent that reacted with proteins), and you got these coloured spots – a fantastic fractionation method compared with anything that had been done before.

GB *You were clearly well aware of advances in fractionation. Was Cambridge a good place to be? Martin and Synge weren't working in Cambridge, were they?*

FS No, they were at the National Institute for Medical Research in London. Nearly all of these fractionation methods were invented by them, particularly Martin. He was the great brain behind this. He was a very bright, original person and a most inspiring person to talk to. I can't say I talked to him. I listened to him, really. I think he was as near to a genius as anyone I have ever met.

GB *Did he come to Cambridge to give a lecture, invited by Chibnall, or did you invite him?*

FS I think I met him at meetings. I don't think I ever heard him give a lecture. Usually someone in his group gave a lecture but he was the brains behind it, a rather eccentric person, and I don't think he did much by himself. He always had a sidekick. Synge was a very good complement to him because he was a very practical, down-to-earth person, and a very efficient experimentalist (Figure 32). Gordon and Consden were the ones who were co-authors with Martin on the paper chromatography papers. Of course, Martin later discovered gas chromatography, which was another great invention.

First PhD student, 1946

GB *You were joined by your first student, was it about this time?*

FS We were joined by Rodney Porter – that was in about 1946. He came at quite an interesting period (Figure 33). He had been in the war and he applied to Chibnall to come to Cambridge. He started

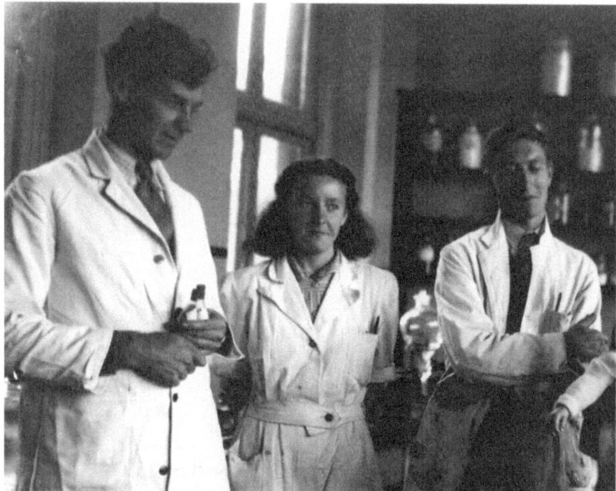

Figure 32. R. L. M Synge (left) and A. T. James (1922–2006) (right) with James' laboratory assistant (centre) at the Lister Laboratories, London, late 1940s. (The Biochemical Society, with permission.)

Figure 33. Fred (right) with Rodney Porter (1917–85) on the front steps of the Department of Biochemistry, Cambridge, 1947. (Courtesy Peter Sanger and Julia Porter.)

Figure 34. Sam Perry (1919–2009), in 1977. (The Biochemical Society, with permission; originally published in Perry SV (1979) *Biochemical Society Transactions* 7: 593–617, copyright holder.)

to do research in Liverpool together with Sam Perry (Figure 34). Both went off to the war; Sam was taken prisoner, Rodney wasn't. He became a major, actually, and then he came to work as my PhD student. He was actually slightly older than me.

GB *How did that work out?*

FS It worked out very well. I mean, he was an understanding sort of person and we became great friends, really. I don't think there was a sort of supervisor–student relationship between us. We were very much colleagues and perhaps I learned as much from him as he learned from me. Particularly, he taught me a lot about research and particularly the light-hearted attitude to research that he had. You should enjoy it, and if things don't work you should jolly well try something else.

GB *He must have been one of your most eminent students because of his work on immunoglobulins.*

FS Yes, I think he was. He was trying to do the end-groups on haemoglobin. We had invented this DNP method and we thought we would apply it to other proteins. It happened that when he was here he read a book by Landsteiner[13] about antibodies and he became very interested in γ-globulin.[14] I told him that was absolutely ridiculous, because we knew there were masses of different antibodies and proteins there and they were quite heterogeneous if you studied it. And so it would be a waste of time and you would get a silly result. He was a bit obstinate and he didn't really respect his supervisor and he tried it out and, lo and behold, he did get a sensible result – just a limited number of end-groups. This went to his head and he kept working on it. It showed for the first time that it was possible to do chemistry on γ-globulins. After his PhD he left for the National Institute for Medical Research, Mill Hill and he kept messing about with these antibodies, trying to fractionate them, to do something with them. For quite a while he did not achieve very much, but suddenly he managed to separate the chains and show the basic structure of the antibody molecule, which of course won him a Nobel Prize in 1972. That opened up a new field completely.

The amino acid sequence of the phenylalanine chain of insulin

GB *Coming back to your work on insulin, you're now starting to sequence. How did it progress? You must have realised you were on to a good thing?*

FS Yes, this was exciting, yes, because it was in fact, I think, the first time any sequence had been done in a protein. I regarded my 1949 paper[15] on the partial digestion of the DNP-protein as more important than the original 1945 fluorodinitrobenzene (FDNB) paper on the end-groups, because it also showed proteins were real chemicals with a defined sequence. I mean we always believed this, but it had never been shown, and there were sceptics who used to say that probably proteins were heterogeneous mixtures. This suggested, really, that if there really were four chains there were two identical pairs.

GB *So you must have been starting to question the four chain hypothesis.*

FS Yes, but we were open-minded. It was quite possible at this stage that it was a symmetrical molecule joined together by disulphide bridges. When we drew the possible structures we usually put four chains into it. We published that in about 1949.

GB *Then you had to embark on the task of doing the whole sequence, without knowing how much there was to sequence, I guess?*

FS Yes, I don't think it went quite as logically as one would like to think. One was messing about doing lots of other things at the same time. We did have a go at seeing if we could just take the chains and partially digest them with acid, and fractionate them on paper chromatograms and get out peptides. We could get out a lot of new peptides. At this stage I was joined by Hans Tuppy (Figure 35, left) and we – he, in particular – started to work on the phenylalanyl chain. Hans was from Austria – this was just after the war and he had had difficulties in getting apparatus to do research, so he was very happy when he came to Cambridge. He had a lot to do. We had the apparatus, and plenty of research projects. He

Figure 35. Hans Tuppy, 1953 (left) and Ted Thompson (1925–2012), 1950s (right). (Courtesy Hans Tuppy and Ted Thompson.)

really got stuck into this phenylalanyl chain. I haven't seen anyone work quite as hard as him. He was only one year with me and he finished the chain, really.[16] We had only done a little bit before.

GB *Was this done entirely using partial acid hydrolysis or were other methods necessary?*

FS No, we got a certain way with acid. We tried to stick to acid as long as possible, but we couldn't get long enough peptides to get a complete sequence. We eventually used enzyme hydrolysis. But we were rather reluctant to use enzyme hydrolysis at first because it was assumed that, being an enzyme reaction, it would be reversible. The Biochemistry lab was of course famous for enzyme kinetics[17] and they all warned us that, if we used enzymes, there would be the possibility of rearrangements. Not only could peptides be broken down, they could also be synthesised. In fact there were theories at that time that normal protein biosynthesis was by the reversal of the action of proteolytic enzymes. We decided in the end to try it out anyway because, obviously, we needed these large fragments to complete the work and we could properly see if there were any rearrangements. We tried pepsin, chymotrypsin and trypsin[18] and we never had any evidence at all of resynthesis. It was possible to complete the sequence from the results with these enzymes.

The sequence of the glycyl chain of insulin

GB *You also had help at this stage of the work. Other collaborators joined you, I understand.*

FS When Hans Tuppy left, Ted Thompson (Figure 35, right) from Australia came to join us as my second PhD student. He was a very different character altogether. He was a typical Australian character.

GB *Did they use the term Aussie, then?*

FS Oh, yes Aussie and 'Fair Dinkum'. He was a very pleasant person to work with, a great joker, and we got on very well together. I had got quite a way with the A chain, but it turned out that the A chain – the glycyl chain – was somewhat more difficult than the phenylalanyl chain. I think the reason for this was that the phenylalanyl chain had several amino acids that just occurred once in the molecule and that made it easier to interpret the results. And there weren't any – not many – unique amino acids in the glycyl chain.

GB *So the overlapping is easier if you have a unique amino acid.*

FS Yes, so we had to get longer sequences, and there weren't any sites for trypsin – so no basic sites in the A chain – so that limited it a bit. We had to use chymotrypsin and pepsin to complete it.[19] We felt at this stage that we had to finish it! There was one amide group that was particularly difficult. You see some of the glutamics (obtained in acid hydrolysis)[20] were glutamic acid and some were glutamines and this was particularly difficult to resolve. We had to get out a peptide and measure its mobility really. Ted stuck to this and eventually got it out.[21]

GB *Proteins can be heterogeneous at glutamine residues; half may have glutamine at a particular residue and half may have glutamic acid, due to deamidation. Was there any evidence for heterogeneity at glutamines in your work on insulin?*

FS No, I don't think there was. But most of our work, you see, was qualitative, so we might not have noticed if a small amount had been deamidated. But we did assume that there was just one structure there. That was a problem with the other peptides and we did get criticised a bit because it was all qualitative work.

Early recognition

GB *Fred, at this stage your work was becoming recognised. You told me you were asked to give popular lectures on how you deduced the sequence of amino acids.*

FS Yes, I used to give quite a lot of lectures – not necessarily popular ones – at meetings. In order to tell the story of the separate chains I used these cards (Figure 36) with the amino acids written on them, and I would lay these out on the bench in front of the lecture and gradually build up the sequence. Maybe we had a dipeptide and then various other dipeptides and I would gradually join them up, and describe how the sequence was built up. This gave rise to

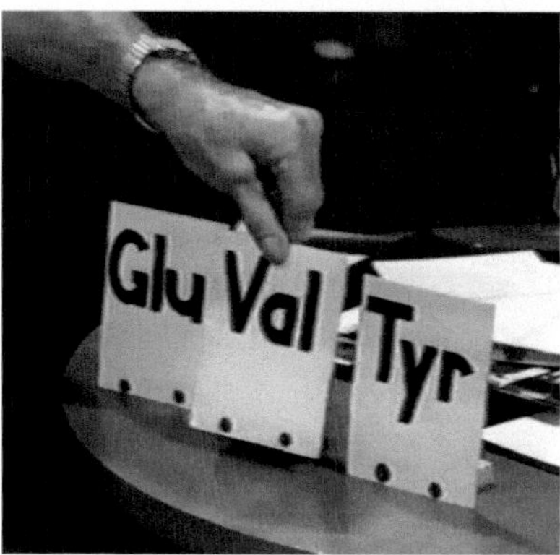

Figure 36. Fred holding three cards symbolising the amino acids, **Glu** (glutamic acid), **Val** (valine) and **Tyr** (tyrosine), when lecturing on the sequence of insulin. (The Biochemical Society, with permission.)

quite a popular lecture and quite a lot of people remarked how much they enjoyed these lectures. It was the first time they had seen amino acids being put together in a consecutive sequence. I too quite enjoyed this. Normally, I don't like lecturing very much. I feel I am not terribly good. I have to take time preparing. With this to play around with, I found it much more fun, explaining how the chain was built up from the peptide sequences, which we had derived.

Arrangement of disulphide bridges in insulin

GB *At this time there was still one remaining problem.*

FS The problem, of course, is that we had these two chains, but we did not know the arrangement of the disulphide bridges. I mean, insulin, on the whole, was a good protein to choose because the chains were short and it was a suitable protein to work out methods. This was the first time amino acid sequences of these lengths had been determined, but obviously we wanted to try and finish the structure, and determine how these disulphide bridges were arranged. That proved particularly difficult. The idea was to try and isolate peptides containing cysteine in partial digests, as a disulphide. They were treated with performic acid to isolate and identify the two halves, thus establishing which pairs were linked. That could be done for two of the disulphides. But the other problem was that there were two cysteine residues adjacent to one another and you couldn't use strong acid hydrolysis to break the bond between them because there was a rearrangement reaction and the disulphide bonds got mixed up.[22] We couldn't find an enzyme that would split between these two cysteines, so there were two possible arrangements of the disuphide groups. By this time the molecular weight (MW) of insulin had been reduced to 6000. That was largely the work of Craig using countercurrent distribution. We were tending towards this value and much preferred that MW because all of our results fitted in with that value.

GB *How did you solve the problem of splitting between the two cysteines, or did you find a way around the problem?*

FS It was just by trying different systems, really. We spent quite a lot of time studying the disulphide interchain reaction which was a pain in the neck, really, because they were rearranging during hydrolysis. But we did find a way eventually of carrying out the digestion in acid, without getting the interchange reaction.

GB *So eventually you got a split between the two cysteines?*

FS Yes, that was largely the work of Leslie Smith. He finally got out a fragment which indicated which way around the disulphide bridges were arranged.[23]

GB *Of course, this is still a problem in protein sequences, especially now that nucleic acid sequences are used to deduce protein sequences. It does not help with disulphides.*

FS That's right, it doesn't help. I think they still use this type of approach – they would not use oxidation but rather reduction.

GB *Were you the first to use oxidation?*

FS Yes, although there had been an earlier paper describing the use of performic acid on proteins which reacted with three amino acids, cysteine, methionine and tryptophan. Fortunately, there was no methionine or tryptophan in insulin. Chibnall and Reese were at that time trying to use reduction methods, but I was not very keen on reduction which did not work very well. Once you reduced the SH groups, they were not very stable and tended to re-oxidise.

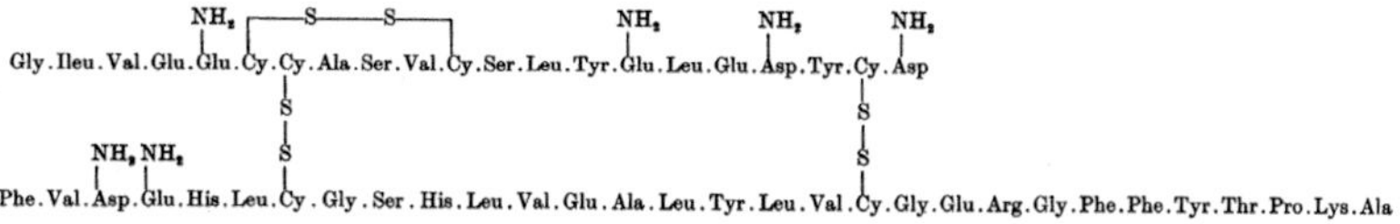

Figure 37. Sequence of 51 amino acids of bovine insulin showing the positions of the three disulphide (S—S) bonds. (Reproduced with permission from Ryle AP, Sanger F, Smith LF, Kitai R (1955). *Biochem. J.* 60: 541–556. Copyright The Biochemical Society.)

The sulphonic acids groups (products of oxidation) were nice and stable.

GB *Particularly with your paper methods, it was good to have a product stable to oxidation.*

FS Yes, and the highly charged sulphonic acid group.

GB *It must have been very exciting when you finished the sequence and structure (Figure 37). What year did you finish?*

FS The final paper was 1955. So it must have been finished in 1954.

First Nobel Prize, 1958

GB *You were awarded a Nobel Prize in 1958 in recognition of this work on sequencing insulin.*

FS Yes, it was a very exciting time (Figures 38–40).

GB *Did you feel like giving up then?*

FS *(laughing)* No, I didn't really!

GB *Some people do!*

FS Well, a lot of people, if they don't give up, take other jobs, become important professors or something like that and usually don't carry on working in the lab.

Active centres of enzymes

FS I was only 40 when I got this Nobel Prize so I did carry on working in the lab. I liked working in the lab and I was never keen to get into administrative work. I wouldn't have been much good at it, I think. So, in fact, it was a stimulus to carry on.

Figure 38. Fred at home at the time of the announcement of his first Nobel Prize, 1958. (Courtesy Peter Sanger and The Biochemical Society, with permission.)

Figure 39. Award of Nobel Prize to Fred in December 1958 by the 'Old' King Gustav of Sweden. (Courtesy Peter Sanger and The Biochemical Society, with permission.)

Figure 40. Son Robin, aged 15, crowns son Peter, aged 12, with a Swedish cap at the Nobel celebrations, December 1958. (Courtesy Peter Sanger and The Biochemical Society, with permission.)

GB *So, you're known for the change you made from the study of proteins to nucleic acids. Was that your intention at that time or did it come upon you gradually?*

FS If I had been asked at that time I would have said I was going to carry on working with insulin. We had the structure. We really wanted to know what it all meant – how it was related to the activity of insulin. I think that was what I intended to do. But we didn't make much progress in that direction. About the only thing we did was to study different species. We found that there were different sequences in different species – e.g. pigs and sheep – which told us something about certain amino acids which were not essential for activity.[24] But we didn't get very far.

GB *No protein engineering was possible then – making point mutations.*

FS No, you couldn't do anything like that! The sort of things you could try was to try acetylating the molecule and see what happens – see which groups are acetylated.

GB *The comparative sequencing obviously gave you some clue as to the important amino acids.*

FS It didn't tell us much because there weren't many differences except for the guinea pig insulin where there were many differences. Leslie Smith did quite a bit of work on that.

But I stuck to proteins for quite a while with the help of colleagues Ieuan Harris, Mike Naughton, Leslie Smith (Figure 41) and others. I was interested in various aspects of proteins. During this period one thing I became interested in was sequencing using radioactive labels. At that time radioactive isotopes were just starting to be used. The main influence then was a visit from Chris Anfinsen (Figure 42). He came as a visitor. He had been using isotopes as a label and he simply got me interested in the idea. One of the first things we tried was to make radioactive insulin.

GB *Were you feeding animals with radioactive ^{35}S-labelled methionine?*

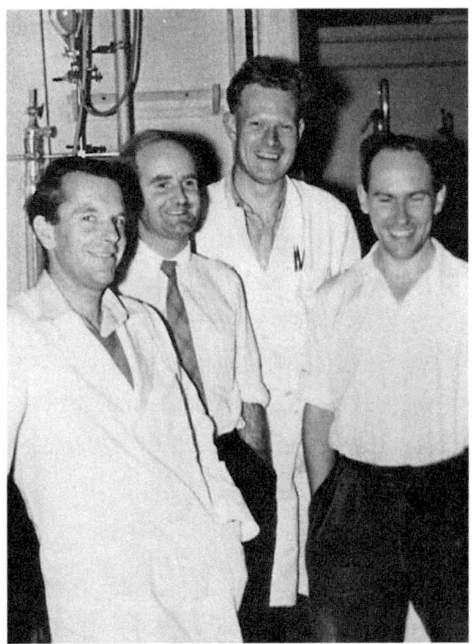

Figure 41. Fred, Ieuan Harris (1924–78), Mike Naughton (1926–2004) and Leslie Smith (left to right) in the Department of Biochemistry, Cambridge, 1956. (The Biochemical Society, with permission.)

Figure 42. Chris Anfinsen (1916–95). (The Biochemical Society, with permission.)

FS Yes, we decided to label a pig! We got a small pig, but we were not allowed to do this in Cambridge. We had to go to Mill Hill. We took this little pig to Mill Hill – to the National Institute for Medical Research at Mill Hill, and injected it there and, I think, isolated its insulin there, using enormous amounts of ^{35}S-labelled methionine. We also, of course, tried with rats. At one stage we tried to label a chicken with ^{32}P-phosphate. And I'm not entirely sure if this story is quite true. We were going to isolate radioactively labelled ovalbumin, which contains phosphate. This chicken laid an egg and the room started to 'scream' – emitting a loud noise. I mean the Geiger counters[25] 'screamed'. We hadn't fully realised there was calcium phosphate in the egg shell! The radiation officer was not pleased.

GB *But although these experiments were not definitive they led on to more specific labelling of enzymes.*

FS Yes, one of the more successful things we did was to try and label enzymes with radioactive phosphate in their active centres. Proteolytic enzymes in particular could be inhibited with DFP (diisopropylfluorophosphate) (Figure 43). If you labelled them up with radioactive DFP, and then did partial hydrolysis, you could then get radioactive peptides.[26] Of course if you did this these

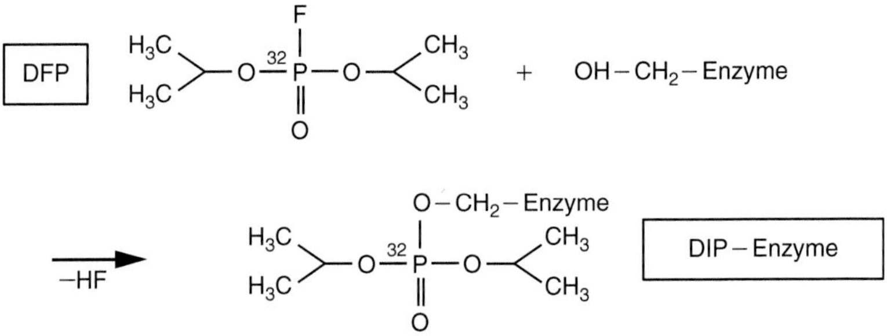

DFP $(CH_3)_2CH{-}O{-}{}^{32}P(=O)(F){-}O{-}CH(CH_3)_2$ + $OH{-}CH_2{-}Enzyme$

$\xrightarrow{-HF}$ $(CH_3)_2CH{-}O{-}{}^{32}P(=O)(O{-}CH_2{-}Enzyme){-}O{-}CH(CH_3)_2$ DIP – Enzyme

Figure 43. Reaction of ^{32}P-labelled DFP (diisopropylfluorophosphate) with an enzyme with a reactive serine at its active centre forming labelled diisopropylphospho-enzyme (DIP-enzyme).

peptides would be contaminated with a lot of other non-radioactive peptides, so the problem was we could not use amino acid hydrolysis to determine their sequence. So we had to make radioautographs (also called autoradiographs) to detect the peptides by their radiation.

GB *Was this the first time you were thinking of a purely radioactive method of determining sequences? This was the technique you used in your subsequent career sequencing both RNA and DNA.*

FS Yes, that's right.

GB *Were you actually successful? It's a difficult problem sequencing proteins by radioactive methods, given that they usually contain up to 20 different amino acids.*

FS On the whole, we were. We developed methods for doing this. For instance you get quite a lot of information from the mobility of a peptide, particularly if you re-hydrolysed it so you could get an idea of the charged amino acids.

GB *Could you distinguish each one of the 20 amino acids?*

FS I don't think you could. You could if you had a marker. For instance, if you had something that was not charged and was smaller than alanine, then you could say that was glycine, and that sort of thing. You could get results from a chromatogram or from paper electrophoresis. Of course paper electrophoresis was used at that time, so you could get information about charge.[27]

GB *I suppose if you knew the sequence of one active centre of a protein, you could compare its radioactive profile with the profile of a different labelled protein.*

FS Yes, you could gradually build up a library of sequences. One of the proteins we determined – this was in collaboration with César Milstein – was a study of phosphoglucomutase, which had a phosphate at its active centre. We did this by these degradation methods.[28] This was also a useful collaboration because he became a very close colleague subsequently.

6

Interview of Fred by the author in 1992: Nucleic acids at the MRC Laboratory of Molecular Biology, Cambridge

Fred Sanger (FS) and I (GB) are seated in a recording studio at Imperial College, London. This is a slightly edited transcript of the final interview.

Move to the new lab

GB *At this time, in 1962, you moved to a new lab in Cambridge.*

FS Yes, that's right. This was to the Laboratory of Molecular Biology (Figure 44).

There were two groups in Cambridge, really, my own group working in Biochemistry and we had nothing to do with the teaching, so we were the poor relations. We were supported by the Medical Research Council (MRC). Then there was Perutz's group in the Cavendish – working in the hut in front of the Cavendish. Both of us were looking for possibilities of new labs and expansion – particularly them. The MRC decided to build a new lab for us – the new Laboratory of Molecular Biology which was completed in 1961.

Figure 44. Laboratory of Molecular Biology, Cambridge, in 2011; it was much smaller in 1962. (Copyright MRC Laboratory of Molecular Biology.) An enlarged MRC Laboratory of Molecular Biology opened nearby in 2013.

GB *You took with you your own extended group.*

FS Yes, the new laboratory had three groups at that time. There was the *Structural Studies* group headed by Perutz, who was also head of the lab. John Kendrew, Aaron Klug and Hugh Huxley were there. And there was Francis Crick's group, called *Molecular Genetics*, later Cell Biology. Sydney Brenner was in that group (Figure 45). There was my group (Figure 46) on the top floor, called *Protein Chemistry*. No mention of nucleic acids at that time. I was joined by various people, who had worked with me in Biochemistry – Leslie Smith who had been on the insulin work (Figure 41), Ieuan Harris (Figure 46) and Brian Hartley (Figure 46). Brian had worked on the important project of sequencing chymotrypsin. He completed that in the Molecular Biology Lab.

GB *César Milstein (Figure 46) also joined you.*

FS That's right. He was in Argentina at the time. He was, again, a colleague from Biochemistry who joined us a bit later.

Figure 45. The Board of the Laboratory of Molecular Biology, 1967. Left to right: Hugh Huxley (1924–2013), John Kendrew (1917–97), Max Perutz (1914–2002), Francis Crick (1916–2004), Fred Sanger and Sydney Brenner. (Copyright MRC Laboratory of Molecular Biology.)

Figure 46. Fred's group leaders in the Protein Chemistry section: Ieuan Harris (1924–78), 1960s, Brian Hartley, about 1970, and César Milstein (1927–2002), about 1985. (Copyright MRC Laboratory of Molecular Biology.)

Sequence studies of RNA in the early 1960s

GB *Fred, shortly after your move to the new Laboratory of Molecular Biology, you started working on RNA. Why was that? Why did you make this change from proteins?*

FS Well, it's quite obvious, I suppose. I don't think there was any sudden decision to work on RNA. You work on something you think you can solve. I think before that time there was no real possibility of working on nucleic acids. The trouble was there was no simple molecule that you could start studying. The simplest RNA viruses were about 3000 residues long, ribosomal RNA was in the thousands and there was no simple RNA. But about that time tRNA was discovered – there were tRNAs for different amino acids, so there was a possibility of having a simple molecule to work on. I think that's mainly the thing that started me thinking about it.

At first it was doing a few experiments on RNA – seeing if we could get any ideas, rather than a complete change. I was still working on protein active centres at the time. This was in the early 1960s (Figure 47).

GB *Which was the first RNA you tried experiments with?*

FS Ah, well. I think probably the main thing that turned us onto RNA was you joining us. I think this was in 1963. You came as a PhD

Figure 47. Fred, 1960. (Courtesy Peter Sanger.)

student and I asked you whether you would like to work on nucleic acids or whether you would like to work on proteins. You said, yes, you would work on RNA. So this rather drove us towards RNA. This was one of the factors.

GB *And you gave me a difficult problem which I didn't solve, I seem to remember. I had to try and purify phenylalanine tRNA.*

FS Yes, this was the first thing to do, wasn't it?

GB *But your first success was using the radioactive methods, which you had developed for proteins, and applying the principle to nucleic acids, wasn't it?*

FS Yes, the first success was about this time, when you were already working with us. Generally speaking sequencing was usually done by column procedures. As I have said before, one of the main technical problems in sequencing is the fractionation of products of partial digests. We had started using paper techniques, whereas other people, for example Moore and Stein,[1] had gone onto columns – particularly ion-exchange columns.

I was always rather keen on using paper techniques – a bit of a prejudice I suppose. But I felt that one could get much more information from a two-dimensional chromatogram than from a column, from which you had to collect fractions and analyse each fraction, which was rather laborious. So we tried to find paper techniques for fractionating small oligonucleotides.[2] In general, they separated pretty badly on normal paper chromatography and paper ionophoresis. But we were able to develop a system that was a two-dimensional fractionation system, and it did in fact give very nice separations. Previously we used to get 'streaks' on bits of paper. This (Figure 48) is an example of a partial hydrolysate of ribosomal RNA – a T1 RNase digest of ribosomal RNA separated in direction 1 by ionophoresis on cellulose acetate and in direction 2 by ionophoresis on DEAE-cellulose. T1 RNase was the main enzyme we used because of its great specificity – only splitting at the Gs. The ribosomal RNA was pre-labelled with ^{32}P by growing the organism, in this case *E. coli*, in the presence of radiolabelled ^{32}P-phosphate. I think this was the first time this technique had been introduced into RNA chemistry. We have continued to use this technique as the main detection procedure. You can make radioautographs and get nice spots of where your substances are present.

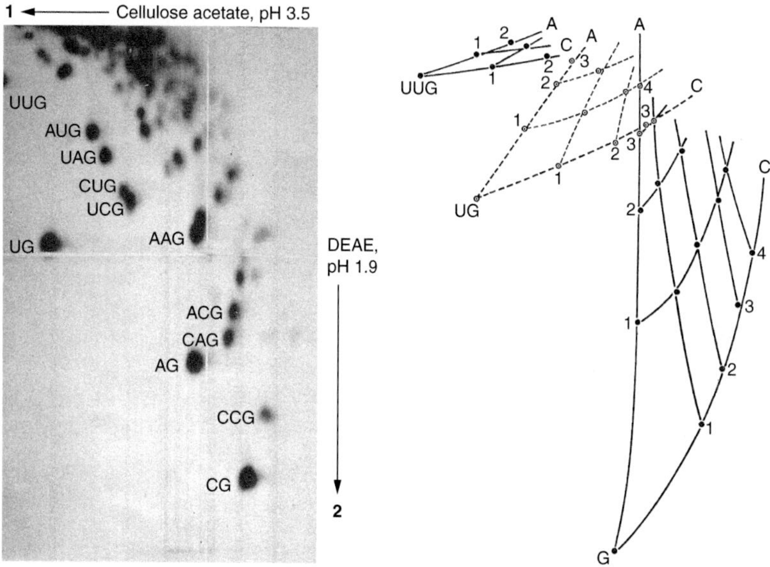

Figure 48. Two-dimensional 'fingerprint'. Left, a T1 ribonuclease digest separated in direction 1 by ionophoresis on cellulose acetate and in direction 2 by ionophoresis on DEAE-cellulose at pH 1.9 of 16S ribosomal RNA. Right, a 'graticule' showing how position defines the composition of an oligonucleotide. (Modified from Sanger F, Brownlee GG, Barrell BG (1965) *J. Mol. Biol.* 13: 373–398.)

GB *How did you derive these sequences?*

FS The thing about this is that we were very keen – as we were with the radioactive proteins – to try to find methods in which you identified sequence from a *position* on a two-dimensional system, or on a one-dimensional system. In fact you can on this system, for small oligonucleotides, because there are only a limited number of nucleotides. We could identify them from their position on the two-dimensional system.

For instance, there are only three possible dinucleotides ending in G, that is, CG, AG or UG, and they are very well separated. There are nine possible trinucleotides and all of them are separated in specific positions on this two-dimensional system. And if you plot this on a picture (Figure 48, right) you get several *graticules*, as we called them. The graticules are based on the number of U residues. So you can identify the composition of the spots simply from the position on this graph. This second graticule is based on one U, and the next with two Us and one G – you can't separate them so well. So we have here a way of getting composition directly from position.

GB *I remember in the first year of my PhD watching you and thinking how remarkable it was that you could separate molecules (isomers) that were so similar chemically but simply differed in sequence. You have written about the fact that when you first saw these separations you were working with Bart Barrell (Figure 49). Even you felt you had achieved something important when you got your first good 'fingerprint'.*

FS Yes, after developing these radioautographs first thing in the morning, Bart brought in this picture with clear spots in it and we got quite excited about that! That was really our first real contribution to RNA sequencing.[3]

GB *So that was published about 1965. But Holley beat you to getting the first RNA sequence.*

FS That's right, but the thing there was that he and his colleagues were the first to get a pure tRNA. He got the alanine tRNA pure. And he more or less applied the classical methods of protein

Figure 49. Fred's assistants: Bart Barrell, 1960s (left) and Alan Coulson, about 1970 (right). (Copyright MRC Laboratory of Molecular Biology.)

chemistry to get partial hydrolysates and they were able to deduce the sequence of alanine tRNA which was the first RNA sequence.[4] So we were beaten to that one.

GB *But I guess you were looking ahead, because radiochemical methods are now used and I was able to help a little with the sequence of 5S RNA (Figure 50) which was a useful model compound at about this time.*

FS Yes, that was 120 nucleotides long.[5] It was really the next one to be sequenced and this method, of course, was very useful for that purpose, although it has no minor bases. tRNA has minor bases and this was an advantage in interpreting the results. 5S didn't.

GB *The minor bases, on the other hand, could cause a problem because in the absence of a marker it was sometimes difficult to be sure what the minor base was. It wasn't directly labelled.*

FS Yes, that's right.

GB *And would you like to tell us how we sequenced the oligonucleotide fragments from the two-dimensional system when we sequenced 5S RNA?*

FS Again, we wanted to use simple methods. The classical way was just to degrade the nucleic acids to oligonucleotides by different methods and then analyse each one. This was laborious. And that makes this analysis extremely complicated because you get so

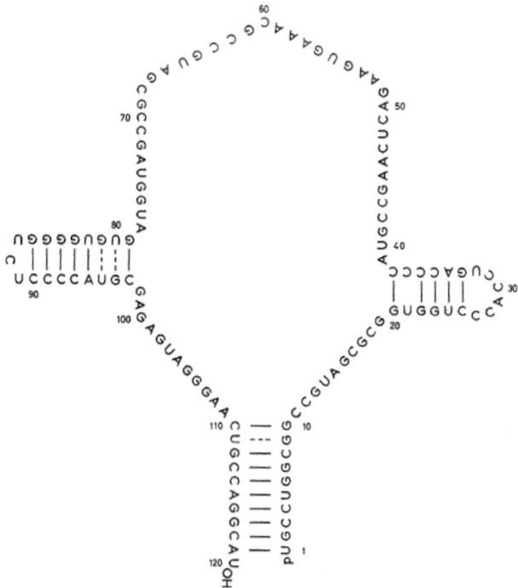

Figure 50. Sequence of 5S ribosomal RNA of *E. coli.* (From Brownlee GG, Sanger F, Barrell BG (1968) *J. Mol. Biol.* 34: 379–412.)

many partial hydrolysis products. So we wanted to find simpler methods of saying what these were. We had these very small fragments, whose sequence we could get simply from their position on the two-dimensional fractionation, but we wanted to have other simpler methods for sequencing the longer oligonucleotide fragments, whose sequence was unknown. So we had to cut them out from the paper and do something with them. The method that was most useful was to treat them with an exonuclease,[6] a nuclease that splits from one end, e.g. with a 5′ exonuclease and then, when applied to a T1 fragment, you will get a mixture of fragments of different lengths all ending in G. So you will get a series of fragments. The first method we used was to measure their relative position on a one-dimensional chromatogram, and that would give us sequences of smaller nucleotides. This method, combined with using other enzymes, was sufficient to work out the sequence of all the T1 RNase products in 5S RNA. But a better method, I think, was proposed

Figure 51. Vic Ling (left) and John Donelson (right), 1970s. (Copyright MRC Laboratory of Molecular Biology.)

by Vic Ling (Figure 51) in his later work on DNA in the early 1970s,[7] which became known as the 'wandering spot' method, where we fractionated partial exonuclease products on a two-dimensional system.

I have an example here with DNA. This is a partial exonuclease digest of a DNA fragment ending in G, so you had to distinguish between C, A and T. There was no internal G residue in this fragment (Figure 52). On this two-dimensional system the relative position of two spots relative to one another is indicated in this vector diagram drawn here. In other words if you have two spots which differ by a C, their relative positions will be indicated by this vector (Fred points). If they differ by a T, it will be like that vector (Fred points, again) so you can deduce the sequence. So if you start here you see the next spot here to the left, so it is a C, the next spot has moved over to the right, so it is a T, and so on, T, A, T, T, A, C. So, one can read off the sequence, CTTATTAC, from the relative positions of the spots on the two-dimensional system.

GB *It wasn't always completely accurate at the ends; one sometimes needed confirmatory evidence, I believe?*

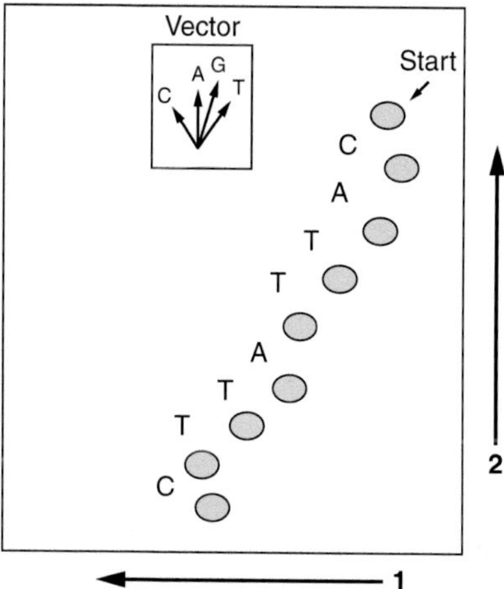

Figure 52. The 'wandering spot' method for sequencing a DNA fragment. (Redrawn from The Biochemical Society, with permission.)

FS Yes, that's right. You could degrade from both ends although the only thing was that A and G were not always distinguishable and that was the main way we determined sequences.

GB *Yes, I remember this method. But it was one of many tricks you developed, too detailed to go into now.*

What about the next stage: with your interest in methods you developed a method called 'homochromatography'. Although we developed this method in the course of the 5S RNA sequencing, it was applied extensively to the larger RNAs, the phage R17 RNA, wasn't it? That was quite a change moving from a small molecule like tRNA or 5S RNA to something that was 3000 residues long. What made it possible to make that jump?

FS That sounded quite a big jump, really. It seemed a bit impossible. R17 RNA was 3000 residues long – it's a bacterial virus and one didn't think it would be possible to simply partially hydrolyse it and get fragments out because of its great size. You would think you would get a hopeless mixture of large fragments if you did a partial T1 digest. But in fact, I think it was Jerry Adams (Figure 53), who thought it was worth a try and he did a partial T1 digest on

Figure 53. Peter Jeppesen, early 1970s (left), and Jerry Adams, 1993 (right). (Copyright MRC Laboratory of Molecular Biology)

the R17 and he got some nice bands that separated well on acrylamide gels. This was quite unexpected but it was correct and there were a few fragments – one in particular – that was quite pure. This we sequenced – I think particularly Adams and Peter Jeppesen (Figure 53) did this work – and it turned out that this fragment had a nucleotide sequence that was related by the genetic code to a known amino acid sequence of the coat protein of the R17 bacteriophage. That was quite exciting because that was the first time that a nucleotide sequence had been determined and shown to be related by the genetic code to a known amino acid sequence in a protein (Figure 54).[8] I mean one of the purposes of going into nucleic acid sequencing was to try and break the code, but in fact the code had already been broken by the time we got there. We weren't the first people to break the code but it was a good confirmation of the code. I think it was worthwhile.

GB *I think we should remember that the genetic code was worked out by some rather clever but indirect methods. I think your work was nice confirmation.*

FS Of course since then there have been many nucleic acid sequences determined and in fact protein sequences are now usually determined from DNA sequences.

GB *You did also, in fact, in collaboration with Bart Barrell, determine tRNA sequences. You have written he was a wizard at managing this and I know he became expert on the minor bases as well.*

FS I think we just got the phenylalanine tRNA pure, and we sequenced it.[9] I don't think we did more than that.

GB *I suppose your work influenced others, such as John Smith on suppressor tRNAs – also Suzanne Cory on methionine tRNA. There*

```
       GA    UGG   CGU   UCG   UAC   UUA   AAU   AUG   GAA   UUA   ACU   AUU   CCA   AUU   UUC   GCU   ACG   AAC   UCC   G
... Ala . Ala . Try . Arg . Ser . Tyr . Leu . Asn . Met . Glu . Leu . Thr . Ile . Pro . Ile . Phe . Ala . Thr . Asn . Ser . Asp ..
    80                                                     90                                                              100
```

Figure 54. The first direct sequence of mRNA: 57 nucleotides of the coat protein of R17 bacteriophage RNA showing the alignment of the mRNA and amino acid sequence. (Reprinted by permission from Macmillan Publishers Ltd: Adams, JM, Jeppesen PGN, Sanger F, Barrell BG (1969) *Nature* 223: 1009–1014.)

was another important thing – you discovered formyl methionine tRNA, didn't you?

FS Yes, well, that was a sort of side issue.[10]

Sequence of DNA: early 1970s

GB *Now comes the most important part of my discussion with you, Fred. It's your work on DNA, which some have described as the ultimate challenge. I suppose you must have had it in your mind when you were working on RNA that you were going to switch to DNA some time. When did you have the confidence, or inclination, to make that switch?*

FS Well, again I think it was a gradual move. The first work in my lab on DNA was by Ken Murray (Figure 55) working as a postdoc. He was doing partial digest on DNA and studying the fractionation of small nucleotides. The problem with DNA at that time was the obvious limitation that there was no way of attacking it because of its very large size. The simplest DNA molecules known then were the single-stranded DNA bacteriophages. The one that featured largely in our work was the φX174 DNA, which had about 5000 nucleotides. It was obviously a bit big to try and start doing sequences on, so it didn't look as though we could apply some of our partial hydrolysis methods to that DNA especially as there weren't suitable enzymes to degrade DNA nice and cleanly – like the T1 RNase that we had used which was very useful for the RNA work. There wasn't anything like that in the DNA field.

GB *So there were two basic problems that put you, and others, off working with DNA. But you found a way around both these problems – the size problem and the specificity problem.*

FS Yes, that was by a copying procedure. We were not the first people to use copying. The first people to copy DNA were Wu (Ray Wu) and Kaiser, and that was a classical piece of work.[11] They were studying the 'sticky' ends of phage λ and they used DNA polymerase. Now DNA polymerase requires a single-stranded DNA template and a primer, which will hybridise to a position on the template, in other words, a piece of single-stranded DNA

Figure 55. Sir Ken Murray (1930–2013) in his garden, 2010. (Courtesy Peter Lachmann.)

followed by a piece of double-stranded DNA. Then it will copy the DNA template from the 3′ end of the primer (Figure 56).

GB *In phage λ, as I understand it, you don't have to put in the primer. It is already there at the end of the DNA.*

FS Yes, you had exactly the right structure with 12 nucleotides sticking out from the end of the λ, which was just the right substrate.

GB *But you realised the implications of that work for sequencing DNA in general.*

Figure 56. A scheme showing the copying of single-stranded DNA by the enzyme DNA polymerase. Copying requires a short DNA primer.

FS Yes, they actually sequenced the end of the λ by quite complicated methods. I think it took them about a year to do it. But that was essentially the method we followed. But the method as described was only applicable to the sticky ends of λ – these 12 nucleotides. We wanted to try and develop a general method whereby we could use the copying by DNA polymerase to get a really short radioactive piece of DNA. If you take a primer on a single-stranded DNA of a phage, then you could make a short piece of radioactive DNA.

GB *But primers had not been used before, and there was a problem in knowing what sequences to make.*

FS Yes, we could figure out a sequence on phage f1, the phage that we were actually using at the time. We could figure out a sequence from the amino acid sequence of the coat protein because there was a sequence that contained only methionine and tryptophan, which have a unique nucleotide sequence. But to make that at that time was difficult because nucleotide synthesis was extremely difficult. It was being pioneered largely in the lab of Khorana (Massachusetts Institute of Technology, MIT) but it was extremely tedious. At a conference I met Hans Kossel, who had worked with Khorana, and he had a similar idea to what we had – to make an oligonucleotide having a sequence complementary to some DNA sequence of a phage and to extend it and make it radioactive. So we decided to get together. He had all the experience of making oligonucleotides, and he made two oligonucleotides and it took them over a year to synthesise the eight residues in each oligonucleotide. But we received them and that was a basis for the next stage of our work.

GB *The first experiment worked, didn't it?*

FS Yes, it worked in that it produced a product, but ultimately we found out it was not copying the piece of DNA that it was expected to copy! But it served our purposes in getting us a nice piece of radioactive DNA to develop the techniques with.

GB *Turning to the second problem, how were you to get specificity of degradation? You had another trick up your sleeve, I believe.*

FS Yes, this was based on the work of Paul Berg (Figure 57). DNA polymerase would not incorporate ribonucleotides into a DNA sequence, normally. But if you substituted magnesium with manganese in the polymerising reaction, then it would incorporate a ribonucleotide. For instance, if you had a mixture of three deoxynucleotides and one ribonucleotide, e.g. ribo C, you could incorporate ribo C into all the positions where a deoxy C would normally be incorporated. The advantage of this is that where ribo C was incorporated, the DNA could be split at the C residues with alkali, or by a suitable enzyme, e.g. pancreatic ribonuclease. So you had a specific method for spitting at Cs. And you could do the same with the other nucleotides, but not so well.

Figure 57. Paul Berg (left) and Walter (Wally) Gilbert (right), who shared the 1980 Nobel Prize in Chemistry with Fred. (Courtesy Walter Gilbert.)

In fact you could do it with ribo G. So that was the first thing. So you had a specific method.

GB *I believe you have something here to demonstrate this.*

FS Yes, this was largely work with John Donelson (Figure 51) and Alan Coulson (Figure 49), in which we would incubate with ribo C and the other three deoxynucleotides, and extend the chain and get a mixture of products all with ribo C extending from the octamer primer. We got fragments which we separated on this two-dimensional system (Figure 58). We used the homochromatography system,[12] described above, which was rather better than the original two-dimensional system for separating larger fragments. Our original method used ion-exchange paper, but this homochromatography method was a displacement chromatography method on ion exchange, thin layers. You actually fractionated the radioactive nucleotides using a 'strong' mixture of non-radioactive nucleotides. So we could separate out these fragments, which were still intact and had ribo

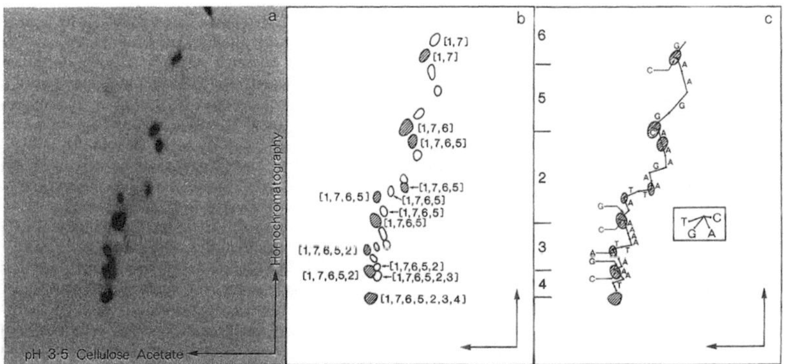

A-C-C-A-T-C-C-A-A-T-A-A-A-T-C-A-T-A-C-A-G-G-C-

A-A-G-G-C-A-A-A-G-A-A-T-T-A-G-C-A-A-A-A-T-T-A-

A-G-C-A-A-T-A-A-A-G-C-C

Figure 58. The first extensive DNA sequence established by primed synthesis with DNA polymerase in 1973: sequence of 58 nucleotides from the f1 bacteriophage DNA. Above, part of the experimental evidence, below the 58-long DNA sequence. (From Sanger F, Donelson JE, Coulson AR *et al.* (1973) *Proc. Natl Acad. Sci. USA* **70**: 1209–1213, with permission of the authors.)

C within their sequence. We could isolate them – they were separated out in a pattern like that (Fred points), and we could cut them out and hydrolyse them with pancreatic ribonuclease – so we got fragments all ending in ribo C. We could determine their sequence by the wandering spot method. So in this way we were able to get a sequence of up to 80 residues in the DNA.[13] That was our first real DNA sequencing method and it was quite an advance on anything that had gone before.

The dideoxy method of DNA sequencing

GB *At this time, I believe you were thinking on the same lines as you had been on the previous work in RNA, and indeed in proteins. You were still using a degradative sequencing method, even though you were using a copying method to generate your substrate for sequencing. But you are best known for your work on the dideoxy sequencing which you published some years later in 1977. Really, this was another jump in your way of thinking. You are no longer thinking of degrading DNA to sequence it. Would you like to explain the method to us?*

FS Yes, this was developed fairly slowly, actually. I'll tell you later about the way we actually developed this technique. But this is the final technique that we came up with, which made use of a copying procedure, again with DNA polymerase, but a copying procedure which made use of competitive inhibitors of DNA polymerase – the dideoxy compounds. These are compounds which look like the normal substrates of DNA polymerase, which are of course the deoxynucleoside triphosphates. But these were the *di*deoxy triphosphates (Figure 59). In other words dideoxyTTP (the first one we used) was the same as dTTP except it didn't have a 3′ hydroxyl group in the 3′ position. But the thing is, if this is incorporated, it can't be extended any further. So, if you have dTTP mixed with a small amount of dideoxyTTP and the other three triphosphates in large amounts, then you will extend the chains, but sometimes the dideoxy will be incorporated and will stop the chain. And if you carry on, all these chains will end at the T residue. So you have a mixture of nucleotides all of which start,

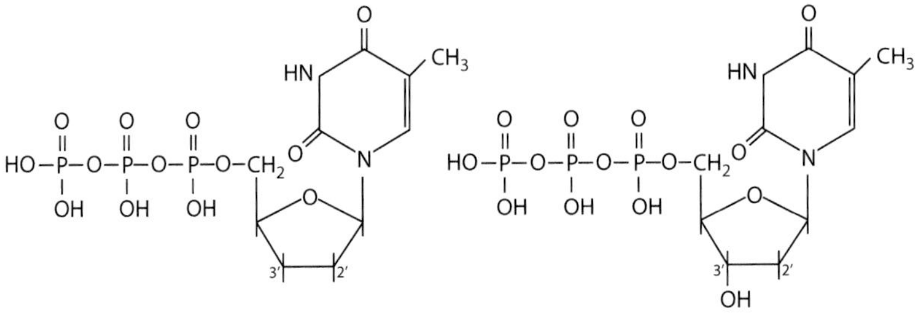

ddTTP dTTP

Figure 59. Formula of 2′, 3′ dideoxyribonucleoside thymidine triphosphate (ddTTP) – a chain terminator – and 2′ deoxyribonucleoside thymidine triphosphate (dTTP). Notice that ddTTP lacks a 3′ OH group present in dTTP, thereby causing chain termination. (Redrawn from The Biochemical Society, with permission.)

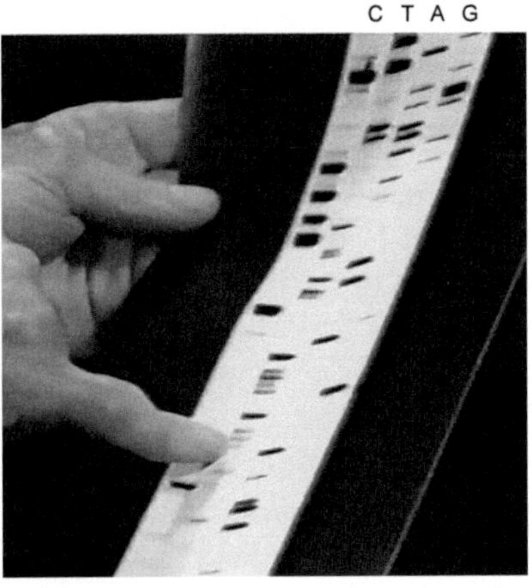

Figure 60. Fred reading off a DNA sequence from an autoradiograph of an acrylamide gel separation of radioactive DNA. (The Biochemical Society, with permission.)

in this case, with the primer – we were using the same system as I have described with a primer before with bacteriophage f1 DNA – and ending with a T residue.

Then you can separate these products and you will get a pattern of bands if you fractionate on denaturing acrylamide gels. This lane (Fred points to Figure 60) is a fractionation of radioactive fragments – all ending in Ts, on an acrylamide gel. You can do the same thing with the other dideoxy derivatives. With dideoxy C you get fragments ending in C, and they will be in different positions on this gel. Because the separation is exactly according to size, one can read off the sequence by looking at the relative positions of these bands on the autoradiograph.

GB *Could you read this off for us now?*

FS Yes, I could. For instance, here we have a nice strong band in the A track, which means there is an A at the 3′ end of that fragment and in that position. The next largest fragment is found here in the T (Fred pointing to Figure 60), so it's T and then there is another T, and another T and then there's a fragment over here, a C, then a T, then a G etc. So, essentially, you can read off the sequence from a one-dimensional reading, and we call this one-dimensional sequencing. Of course it's very much faster than any of those degradation procedures, which we used previously. We don't have to cut out these bands or do anything with them. We know the sequence straight away from the fractionation procedure.

GB *Now, we all know you didn't dream this method in the bath. There were preliminaries between the early 70s, where you have described how you started working on DNA and solved some of the problems, and this beautiful read-off method with its simplicity and speed. Could we discuss the developments which made this advance possible? It seems to me they are in several parts. There's the part relating to how you came to use the terminator. Then the parts related to the gel technology. And there may be other parts. How did you come, in particular, to use the dideoxys. Were there preliminaries to it?*

FS Yes, it's a bit difficult to know just how it started. It was of course a completely different principle from before. You know, I've been trying to look through my notebooks recently to see how I did develop it. I had made a suggestion as to how it might have developed – written, and in lectures I had given, but I now find I was wrong in this and looking through my notes it was rather by a roundabout method that I came on this. The thing is that at the time we changed over to this one-dimensional method we were using this ribo C method and fractionating fragments on two-dimensional systems. At one time we got fragments out using this and we tried to extend the use of what we called the wandering spot to this method. For that, one had to produce very complex mixtures of nucleotides to get every fragment. So we spent quite a lot of time trying to get a really complex mixture of products from a priming experiment of this sort. Some of my earlier experiments were simply doing that from the point of view of doing the ribo C and degradation, and using the wandering spot to study them. You get a sort of sequence here but you want to join it all up. But it occurred to me that if one *could* get fragments all ending in one particular residue then you could get more information from these experiments.

For instance you could get fragments ending in A. This was based largely on the work of Paul Englund,[14] who had shown that if you take a mixture of fragments and treat them with T4 DNA polymerase, a normal synthesising polymerase but which also has a 3′ exonuclease activity, if you only had one nucleotide triphosphate present, say ATP, then it would degrade these fragments until it comes to an A. So you could get a mixture of fragments all ending in A. So that's how we first came across the idea of getting fragments all ending in A using this method with T4 polymerase. That's what we call the plus (+) method. But the first experiment we did was just to try and fractionate by the homochromatography two-dimensional system and that wasn't terribly good, but it looked slightly hopeful! So that encouraged us to look at this further. We then tried another system, called the minus (–) system, which was actually just three nucleotides – missing out one nucleoside triphosphate, e.g. TTP. So there were three triphosphates, so that if we didn't have T there, all the fragments would be extended to just before a T. So that was another way of getting all fragments ending in the same residue. That worked quite well with *E. coli* DNA polymerase.

In my notes I found the first experiment I did in early 1973 using the plus system and then I used the minus system. It looked as if I had a snippet in my notebook earlier in June 1972, in which I had made a little diagram, which I think summarised probably the plus method. I was either perhaps figuring it out myself and thinking if I could do this, or perhaps explaining it to someone else.

GB *Does this post-date the publication of the Englund paper?*

FS I think it was about the same time.

GB *So this account gives an insight into the development of the plus and minus method, but you were not satisfied with this were you? You have written you were not satisfied with the quality of the data.*

FS No, using the plus and minus system[15] was a great improvement on what had gone before and one could read off sequences but the bands were in very different concentrations and very often many of them were missing from either the plus or minus system. Nevertheless, you could get 50 residues off a single experiment which was a great advance on anything that had gone before. We applied this to φX DNA and did the whole sequence except for just a few errors.[16]

GB *I think that is the first paper I ever read where authors admitted there might be errors in their sequence!*

FS It was quite a big sequence! And I think we knew (Figure 61) there were a few errors.

GB *But that was not the only thing you developed, was it? An equally important development was the gel technology. After all, the work using homochromatography on a thin layer was very different from*

Figure 61. Fred at the bench, 1970s (left); Fred seated with Bart Barrell, about 1980 (right). (Courtesy Peter Sanger.)

the gel technology. Can you tell us how the gel technology developed over the years?

FS Yes, acrylamide gels were a fairly standard thing for separating proteins etc. At the time we started out John Donelson (Figure 51) was doing some experiments with acrylamide gels. We were largely using homochromatography because we thought that was rather better for the larger oligonucleotides, but I gave some of these samples to him and he ran them on acrylamide gels and they did look marginally better than the homochromatography. They weren't very good at the time but they were run on much smaller gels than we would now normally use – yes smaller gels, and not under denaturing conditions. We then started to use much longer gels under denaturing conditions.

GB *You would have had to have a glass-blower make longer plates for you. You couldn't buy these things then, could you?*

FS Yes, or you made them yourself. They had to be nice flat plates. You could make a fairly thin gel and that very much improved the resolution.[17] The unexpected thing about this is that you could find a method that could fractionate fragments absolutely according to size. And it depends on that. I think when we first started doing this it was just another crazy idea. How could we possibly find a system that would separate things exactly according to size? So we didn't take it too seriously, and I didn't make careful notes when we started on it. But in fact you can separate things according to size on acrylamide gels. It was not too easy at first. On the first gels we found that we got very bad inversions and sometimes the larger fragments were running faster than the small ones. And that completely upset things.

GB *Do you remember, did you try different pHs, initially?*

FS I think we tried most things, yes. I mean highly denaturing and a good concentration of urea were what we finally got to. And another thing was to run them at high voltage, so that they got pretty warm; and check that the DNA was absolutely denatured which was really important, I think.

GB *Another thing that is absolutely critical to the development of the dideoxy sequencing method was a supply of the dideoxynucleoside triphosphates. Only dideoxyTTP had previously been synthesised because Kornberg had worked with it many years before. How did you get into trying the dideoxy triphosphates as terminators?*

FS Yes, well I think with the plus and minus method one could see if you could get a terminator inhibitor that would be another possibility for reading off the sequence directly from the gel. There was no reason why it would be any better than the plus and minus method. But I looked around to see if I could find any. Luckily, I came into contact with Dr Klaus Geider, who had made some dideoxyTTP. And he let us have a bit. Alan Coulson ran it on one of our samples and it gave a beautiful pattern. I mean bands all the way up the gel and they were sharp. That was very exciting. So it looked as though that would be much better. The other three dideoxys had never been made and were not available anywhere. So Alan Coulson and I had to set down and just prepare them! This was something which we had really no experience of – nucleotide synthesis. Luckily, we had Mike Gait in the lab – an expert in nucleotide synthesis, and Bob Shepherd and they were very helpful to us, and I had a certain amount of experience of chemical synthesis. But nucleotide synthesis is rather specialised. It's not quite such fun as making dyestuffs and things where you get beautiful crystals. You can never crystallise these nucleotides. You just separate them on some type of chromatogram and see how pure they are.

GB *This must have been a considerable risk for you as, I understand, it took you about a year to synthesise these dideoxys?*

FS Yes, you sort of learn the methods and of course they had never been made before. We used a method analogous to the one used for dideoxyTTP, as far as I can remember, but there were difficulties. But we did manage to get them in the end.

GB *Were they pure when you got them – or was that not the consideration? Did they just have to give a good picture on a gel?*

FS That was the test, but I think we did get them pure.

GB *Using paper methods?*

FS I'm not sure! I think we used some columns. It was just straight chemistry.

GB *Fred, would you like to tell me something about the further developments in the field of sequencing after you published this work? Because your work was on a single-stranded bacteriophage*

DNA, it was not immediately applicable to the usual way DNA comes – as double-stranded DNA, was it?

FS Of course, there have been further developments. One thing was the method of Maxam and Gilbert[18] which was similar in the result you get, but it used a partial chemical hydrolysis method for obtaining the fragments. This was done by having a reagent that split relatively specifically at a given residue – although not 100%. But the Maxam and Gilbert method worked very well. It was developed about the same time as the dideoxy method. It came after the plus and minus method and was used probably in preference to the plus and minus method in the States. It was a very ingenious method. And it employed these acrylamide gels again and so you got the same sort of pictures. But the thing was the reactions they used were not specific and it occurred to me if you used such reactions you would not get sensible results. But it turned out that it didn't matter. If you don't get the reaction going specifically, actually you have more data. If you get the G going quantitatively, and the A going half – a faint spot for A and a strong spot for G, then you have actually more data.

GB *Fred, I was going to say there was a friendly rivalry between those using your method and those using the Maxam and Gilbert method. In fact you shared the Nobel prize with Gilbert and Paul Berg (Figure 57). So both contributions were recognised.*

FS Yes.

The second Nobel Prize

GB *You were awarded* two *Nobel Prizes. This is most unusual. Was it easier to get the second one than the first one?*

FS I think it was really. When you've got a Nobel Prize you get good facilities for research, you get good students, excellent postdocs who used to come to the lab and work with us. And another thing, having had a Nobel Prize, I didn't feel there was any obligation to get papers published. I could spend my time doing rather crazy 'way-out' experiments. I think this is how you get important advances very often, by trying the way-out experiments that are very likely not to work – crazy experiments. I did lots of experiments that didn't work. I enjoyed it more and kept at the work.

GB *Excuse me, but do you think scientists are too conservative in the experiments they try out?*

FS Well, I think they have to be very often. When you're obliged to get papers and get your name in the literature – keep your name near the top of the list, you feel that you can't really try crazy experiments. But I was privileged in a way, because at first I didn't have any financial problems and, after 1958, I was in a position in which I didn't really have to publish papers or get grants and things. I had a permanent job. And I could do such experiments and I felt obliged to do such experiments. I think that helped me very much in the development of the DNA sequencing procedures.

GB *No doubt you felt pleased to have your work, and that of your collaborators, acknowledged by the award of the second Nobel Prize. I believe you have brought along a photograph of your presentation.*

FS Yes, this is the occasion in 1980 where I was presented with the Prize by the present King of Sweden (Figure 62).

GB *And you would have met the other laureates, Wally Gilbert, there if not before?*

FS Yes, I think I saw quite a lot of him at this occasion in 1980, and his family too, because these occasions are quite exciting and you can take your family along and you get treated more or less as royalty. When you are at home you're just an ordinary scientist. Nobody takes much notice of you. When you're that week in Sweden you are really fêted and it is a very exciting occasion. This is me with one of the princesses (Figure 63).

GB *One gets to practise one's dancing skills?*

FS A little bit, yes. The main thing about it is, if you have your work recognised in this way, you do feel very gratified to know your work is appreciated and it gives you confidence to carry on.

Early cloning work with single-stranded bacteriophages

GB *But in your case you started a ball rolling – sequencing – that has a momentum of its own now. We've discussed some of the future*

Figure 62. Nobel Prize presentation by King Gustav Adolf in Stockholm, 1980. (Courtesy Peter Sanger and The Biochemical Society, with permission.)

developments but what about the early cloning work in single-stranded bacteriophages?

FS One of the disadvantages of the dideoxy method was that it was only really applicable to a single-stranded DNA – you had to have a single-stranded template but that limited one very much because most DNA is double-stranded. So how could one get around this? We tried several different methods at the time. But the answer, I think, came from the work of Messing. Messing and Gronenborn[19] developed a method for cloning double-stranded DNA into a single-stranded DNA bacteriophage – M13. That made it possible to sequence pieces of double-stranded DNA. You put them in the cloning vector and then used a primer for sequencing them that was part of the vector – a flanking primer. You could therefore use the same primer on every cloned fragment.

GB *I know you credit Messing for this method, but – correct me if I'm wrong – it was your idea to use a universal primer, was it not? And*

Figure 63. Fred with the King's sister at the banquet in City Hall, Stockholm, 1980. (Courtesy Peter Sanger and The Biochemical Society, with permission.)

was it not your idea to randomly degrade the DNA before cloning? I thought these were contributions from your lab?

FS We developed it quite a lot, but I think the credit must go to Messing really, because he was the first person to make the vector and actually do it. I think quite a lot of people had thought about it but he was the first person to do it. It made the dideoxy method universal, really, and usable on any system. It is the method that is now being used.

GB *My next question is now that a huge mass of data can be produced by one scientist, or one technician, can you tell us about the early use of computers to help organise the data? I think you did collaborate with people, such as Rodger Staden, to help organise data. This was another quite interesting aspect of the work.*

FS Yes, when we were starting on this, I was a bit rude about computers. People used to come and say to me isn't it about time you got onto computers, and started to put things in. But we didn't really need computers when we just had a sequence of 50 residues. Part of the fun was looking at the radioautographs and figuring out what the sequence was! You could write it down quite simply. But certainly as things have gone on, methods have become much more automated and now results come out at a terrific speed, and so one very much depends on computers and rather complicated computer programs. I have not been instrumental in anything to do with computers. The person who was helping us largely was Rodger Staden (Figure 64), who was a computer expert attached to the X-ray crystallography group in the laboratory, and he formed a friendship with Bart Barrell and helped us very much to develop programs. Actually the first program we had was brought in by Mike Smith (Figure 64). He was working with computers – I think his brother was a computer expert – that was the first program we had, and that was just for storing the data with φX DNA, which incidentally we eventually completed. It was the first DNA to be sequenced, really.[20]

Figure 64. Rodger Staden, 1975 (left) and Mike Smith (1932–2000), Nobel Laureate, 1993 (right). (Copyright MRC Laboratory of Molecular Biology.)

The progress of sequencing over the years

Year	Protein	RNA	DNA	Number of residues
1935	Insulin			1
1945	Insulin			2
1947	Gramicidin S			5
1949	Insulin			9
1955	Insulin			51
1960	Ribonuclease			124
1965		Ala tRNA		75
1967		5S RNA		120
1968			Bacteriophage λ	12
1978			Bacteriophage φX174	5 386
1981			Human mitochondria	16 569
1982			Bacteriophage λ	48 502
1984			Epstein–Barr virus	172 282

Figure 65. Progress of sequencing over the years: a table of selected proteins, RNA and DNA sequenced from 1935 to 1984. (Redrawn and modified from Sanger F (1988) *Ann. Rev. Biochem.* 57: 1–28.)

GB *We have a table here (Figure 65) of the various stages in the development of sequencing.*

FS This table shows how from the first work on insulin in 1935, where the first amino acid was located in insulin. That was not our work; it was the work of Jensen and Evans,[21] who actually showed phenylalanine was at the beginning of insulin. Then in 1945 we had two residues sequenced in a protein, and that was our glycine and phenylalanine. In 1947, the Martin group then sequenced gramicidin S – five amino acids,[22] and then in 1949, we did the partial digestion – nine amino acids sequenced of insulin, and by 1955 we had 51 of insulin. The next stage, an important landmark in protein sequencing, was ribonuclease, which was 124 amino acids – that was the work of Moore and Stein, using largely ion-exchange columns.

Then we go onto the RNA. The alanine tRNA – 75 nucleotides, that was the work of Holley, and 5S RNA – your work, when it was the largest RNA sequenced. We then can go on to the DNA, and the first piece of DNA was the bacteriophage λ – 12 nucleotides, the work of Wu and Kaiser, then the next one – a big jump up to the φX174 in 1978. Then we worked for a while on mitochondrial DNA, that was human mitochondria, and that was the first thing we did completely with the dideoxy method, in the

16 000 range – again the biggest that was ever done. That was really the first bit of *human* DNA to be sequenced and we can say we were already doing the human genome at that time.

After the dideoxy work, Alan Coulson and I did the phage λ – 48 000; it was something of an anticlimax. But we managed to complete that, all by the dideoxy method and using the cloning procedure of Messing. I think the next one was the Epstein–Barr virus and I think we could sequence longer DNA than that now, though there's probably no longer DNA sequenced as yet. But there are already millions of sequences in the data base now.

GB *I think it is remarkable what has happened even in the eight years since your retirement. But that takes me on to ask you what has happened since you retired. I know you said your last major work was the work on bacteriophage λ; and with characteristic modesty you said that was not very important. It* was *important because it showed that substantial pieces of DNA could be sequenced.*

Sequencing genomes

GB *So much has happened since you retired. People are discussing sequencing the whole human genome. Large parts of the* E. coli *genome and the* Caenorhabditis elegans *genome are now being actively sequenced. I think I, and others reading this, will be interested in your view about the desirability of sequencing the human genome. Dare I ask you how long you think it might take?*

FS Well I don't want to look into the future very much. I'm a scientist not an astrologer! I don't like to predict things you can't be too sure about.

GB *But are you in favour of the initiative?*

FS Yes, yes, I think it's a logical conclusion of the work I've been doing really. I think 50 years ago I felt sequences were the important thing. I have spent most of my life trying to develop methods for sequencing. And of course the DNA sequence is the ultimate – you can get the sequence of the RNA and protein from the DNA. So I feel that it is a very worthwhile project. And from the medical point of view it is likely to prove useful.

GB *Had you not decided to retire would you have worked on the Human Genome Project?*

FS I think I might have. Alan Coulson (Figure 66), who was working very close to me for about ten years or so, has now joined John Sulston (Figure 66) and is working on the nematode genome. This had become a big project, really. It's well into the millions, and they have been developing initially the mapping procedures and are now on to sequencing using four of these Applied Biosystems automatic set-ups. The whole method has been automated now to a very great extent so you can turn out these sequences at a very great pace.

GB *It occurs to me that Applied Biosystems have done to nucleic acid sequencing what Moore and Stein did to your paper methods for sequencing insulin.*

FS It's a bit different, I think. It's using our chemistry, and the acrylamide gels, just the same, but automating it.

GB *You get peaks, so it makes it more quantitative. Moore and Stein were interested in quantitation, so I've read.*

FS But it isn't a quantitative problem, is it? It's a qualitative problem.

Figure 66. Alan Coulson, 1987 (left) and Sir John Sulston, about 1985 (right). (Copyright MRC Laboratory of Molecular Biology.)

GB *But when you get to the high resolution region of the gels, when you're trying to read 800 nucleotides, it can be a quantitative problem, because don't they run editing programs on the data which couldn't be done if the data weren't quantitative?*

FS Yes, that's true.

Overlapping genes

GB *You did discover, perhaps almost unexpectedly, an interesting property of the φX174 genome that was* not *predicted. I am referring to the overlapping genes paper.*[23]

FS Yes, that was an exciting thing, really. We didn't really expect it. We've only really summarised the methods for developing sequencing so far. During our work I was doing lots of other experiments – trying to go off in different directions, and of course, also, we got results. And these were quite interesting. Particularly in the φX work, we found several examples where the DNA could be read in two phases giving rise to two different amino acid sequences.

GB *And this is also fundamental. The idea that a piece of sequence could lead to two frames was rejected by the early theoreticians.*

FS I don't know if it was ever really considered. The classical theory was 'one gene, one enzyme'. One DNA sequence, one protein. But, in fact, you can have two genes overlapping one another in the φX. Now it's not a very general thing. I think in higher organisms it's probably unlikely because, in fact, unlike the φX where you have to cram as much as you can into one small DNA sequence, in higher organisms it seems as though you have much more space – much more DNA, than you really require for the amino acid sequence. But it was a very interesting thing.

Mitochondrial DNA and tRNA sequences

GB *One of your other surprises in the sequencing of the mitochondrial DNA, in collaboration with Bart Barrell, was quite unexpected, wasn't it?*

FS Yes, well, mitochondrial DNA was the DNA we did after the φX. We set about trying to do human mitochondrial DNA. And, what we found was, we knew certain protein sequences, we found a lot of sequences which appeared to be coding for proteins but contained terminating codons – UGA. And this was a bit of a mystery, because this is a terminator! You couldn't have a protein with UGA in it because the genetic code was considered to be universal, and UGA was a terminator! And this was a big puzzle but eventually when we put a whole lot of sequences together and using amino acid sequence, as well as nucleotide sequence, it was clear that this UGA was *not* a terminator. It was actually a codon for tryptophan. This was the first time that the genetic code, which was thought to be universal, was shown not to be universal. Mitochondria had a different genetic code from the normal genetic code used in most bacteria and other mammalian organisms.[24] And, we found a few other differences in mitochondria[25] and differences between the yeast mitochondria and the human mitochondria.[26]

GB *And this, of course, led on to looking at the tRNAs that were specific for the mitochondria, didn't it? I believe Bart Barrell made contributions in that area.*

FS Well, from the mitochondrial DNA sequence. You see these were coding for certain tRNAs. And Bart was excellent at looking for mitochondrial tRNAs by analogy with what one expected for a tRNA sequence. It turned out these tRNAs mostly were in between protein coding genes.

The new Sanger Institute in Cambridge

GB *I believe a new institute is being set up in Cambridge to look at the human genome. Would you like to tell us about that?*

FS Well this is – it's still on the drawing board, I think. The idea is to set up an institute headed by John Sulston. He has been working with Alan Coulson on the nematode genome and the idea is to extend this considerably, initially to finish the nematode genome and then to work on the human genome. Of course many people are working on the human genome. This is going to be very much an international effort, I think. There is collaboration between the groups. There is a lot going on in America in various laboratories. But, you know, there is plenty to do there and I think it is an exciting development already. People are beginning to find out things about hereditary diseases, where applications are more likely to come about. I am gratified to see this extension of my work continuing on an unimaginable scale.

GB *Have you got your speech ready, because I believe this institute is to be called the Sanger Institute and no doubt you will be asked to open it?*

FS (Laughs)

Retirement, 1983

GB *You retired, I believe, in the early 80s.*

FS Yes, I retired then, when I became 65, for various reasons, but I believe it's the normal retiring age in this country. So I decided I would retire. My work had sort of come to a climax with the development of the dideoxy sequencing method. And I felt it would not be so exciting to go on working. Also, I think I did not want to do what most people do – take up administration, keep on a lab, and just potter around, really, not doing full-time research. I would either have wanted to do full-time research in the lab, or retire. Particularly the memory gets a little less brisk than it was and I decided probably I couldn't make very important discoveries any more, and I thoroughly enjoyed retirement.

GB *It has allowed you to pursue some of your hobbies, no doubt?*

FS Yes, yes. It turns out I haven't done all the things I intended to do. I've been mainly involved in gardening which I spend a lot of time on (Figure 67).

GB *Did your family life affect your scientific career?*

FS Yes, I certainly think it did, yes. I haven't said much about my family life (Figure 68), but essentially it was a very quiet and peaceful and happy time. I married Joan Howe, no problems there. I think if people have these sort of problems – don't get on with their wife, etc., – then you spend an awful lot of time thinking about those sort of things. If you're going to be a scientist, really

Figure 67. Fred in his rose garden at his home, 'Far Leys', in retirement. (Courtesy Peter Sanger.)

Figure 68. Fred and Joan Sanger in retirement at their home, 'Far Leys', 1980s. (Courtesy Peter Sanger.)

Figure 69. Fred and Joan's children: Robin Sanger, 1983 (top), Peter Sanger, 1991 (centre), Sally Sanger, 1982 (bottom). (Courtesy Peter Sanger.)

Figure 70. Fred with his grandchildren on a new river boat at Ely. (Courtesy Peter Sanger.)

Figure 71. Fred boating in the Medway estuary. (Courtesy Peter Sanger.)

you have to live with the problems. Make all your problems the scientific ones! And, you know, with a quiet family life and a happy family life one could have time to contemplate nucleic acids and proteins.

I think I owe a great deal to my family. I have not only a wife, but I have three children (grown up now), Robin, Peter and Sally (Figure 69). Two boys and a girl and we used to have exciting holidays together. And I took fairly long holidays on the whole.

GB *Did you sail with them? Boating is one of your hobbies.*

FS Yes, I suppose really it was my second son, Peter, who started me off on this boating craze. He became very keen on boats and he learnt to sail. He went to Bryanston – both the boys went to Bryanston. He learnt sailing there. He taught me to sail. In 1960, I think, I got my first boat which was on the River Medway. I spent a lot of holidays on that. We used to do crazy things. We used to take it out to sea down the Cam and then out into the Wash and it wasn't really a seagoing boat. It was a flat-bottomed thing.

GB *I shouldn't have come! You once invited me.*

FS Yes, it could be fairly hairy!

GB *So, you were a good father, but you're now a grandfather?*

FS Yes, my second son, Peter, has two sons and they are getting on now. The other two haven't produced anything yet. But this sailing has been something I have gone in for as a leisure activity. After this first boat we had a number of sailing boats and I partially built a sailing boat which we kept on the sea, and that was quite an exciting experience. That is to say, when I retired, I bought a hull and partially fixed up the inside of the hull. This (Figure 70) was a motor boat I acquired after I retired, because I got a bit old to handle sailing boats by myself (Figure 71). And we keep that on the river and we spend quite a lot of time in it. This was a sort of return to my childhood where we used to do carpentry as a boy and we got a lot fun out of building and fitting up this boat.

GB *Fred, did your children follow you into science – into biochemistry, in particular?*

FS No, no, they didn't, not at all. Whether I pressurised them I don't know. I didn't intentionally, but I think I might have done! Especially with the elder one – he was the bright one. I think, you know, if you've seen your father do a job, it's not so exciting. None of them actually became scientists. The elder one, Robin, he studied law. He's quite bright. He became a bank manager. The second one, Peter, he wanted to study medicine, he would have liked to study medicine, but he wasn't academic. He didn't manage to get through. He works at the tax office. Sally, the youngest girl, was more academic and she went to Girton, Cambridge. She studied English and started to do a PhD, and spent quite a lot of time on that, but so far hasn't finished it. She says she might finish it. She's gunning for information and technology and works in a library, essentially. And she's the more academic one. None of them are in science.

Looking back

GB *Finally, Fred, would you talk more generally – not so much about science. What do you think are the most important things, in general, that allowed you to make these huge advances in molecular biology from when you started in the early 1940s to now? To me, looking at this period, it is amazing what you, in particular, have achieved.*

FS I think it's partly a question of being at the right place at the right time. It is an extremely exciting period in biochemistry. Apart from my own work there have been terrific advances going on during the last 50 years and I think it's the most exciting field to be in, and the most exciting time to be in the biochemical field. It's very difficult to say what helped but I certainly have been lucky in being able to be independent – starting off firstly with some of my own money to help me with the early stages of my work, and I haven't mentioned before that I was supported for seven years by a Beit Fellowship, which was a fairly generous fellowship in those days. For seven years I was supported by them while I was doing the insulin work. And of course then I had the first Nobel Prize at a rather young age, and I had very good facilities for doing the nucleic acid work. Also, I have been among very helpful collaborators, in both labs where I have worked. There has always been a helpful co-operative attitude towards people and there have always been people to talk to who had lots of ideas. Maybe I've had a few ideas but I never know where they come from . . . they may come from talking to other people around the lab. And I think one owes a great deal to one's colleagues and other people who are around. Very often these ideas come from one's students. We had very good people coming to the lab – kept us young really.

7

Post-Sanger sequencing: high-throughput automated sequencing

After Fred Sanger described his dideoxy sequencing method in 1977, the main technical innovation was the development of high-throughput automated methods that were not dependent on radioactivity. Autoradiography was replaced by the introduction of fluorescent 'tags' on short DNA 'primers' – one for each of the four dideoxy terminator reactions. The four dideoxy terminators – C, A, G and T – were distinguishable by the four different fluorescent tags on the primer used for the particular dideoxy terminator sequencing reaction.[1] These tags were detected by lasers and a camera operating in real time monitoring the progress of the separation of the chain terminated reaction products during their separation by gel electrophoresis. The sequence was now recorded automatically. The raw sequence information was processed by suitable software programs on a computer and presented as a linear sequence. Thus Fred Sanger's manual 'read' of radioactive bands on an autoradiograph of a gel was now replaced by an automated 'read' of fluorescent bands on the same type of gel.

Commercial machines using this technology soon became available and were first used by Craig Venter and colleagues in 1987.[2] Subsequently there were further technical improvements to these machines optimising the matrix used for the electrophoretic separation of the chain terminated reaction products. Improvements were also made by positioning the fluorescent tag on the dideoxy terminating nucleotide.

But these technical developments were still using the basic dideoxy sequencing method, the copying of a single-stranded template by primed synthesis with a DNA polymerase in the presence of dideoxy chain terminators, as devised by Fred Sanger in 1977.

Nevertheless, these improvements were vital. It was now possible to consider scaling up to sequence the human genome, by creating a 'sequencing factory' type of operation. Automation was obviously essential for tackling the sequence of the human genome because of its large size. The human genome is about 3×10^9 (3 000 000 000) base pairs whereas the single-stranded bacteriophage φX DNA sequenced by Fred Sanger's team in 1982 was a mere 5386 bases long. This is exactly what happened in the 1990s.

A critical decision was made by the Wellcome Trust (a UK charity) and the UK Medical Research Council in 1993 to fund the efforts to sequence the nematode worm *Caenorhabditis elegans* (introduced into science by Sydney Brenner as a relatively simple model multicellular organism) and the human genome in the UK at a new laboratory outside Cambridge in the small village of Hinxton. John Sulston and Alan Coulson, who had started the nematode genome project at the Laboratory of Molecular Biology, Cambridge, moved to Hinxton Hall along with Bart Barrell and many other scientists, including David Bentley from Guy's Hospital, London, who had experience with human genetic disease. A refurbished building was opened in laboratories bought from a metallurgy and engineering group, Tube Investments, with John Sulston as the first Director. Fred Sanger opened this new facility in October 1993 arguing in his speech that it was now timely for an attack to be made on the ultimate sequencing problem – the human genome. He went on to say that the methods for doing this through genome mapping and dideoxy sequencing were in place. Characteristically, he highlighted the contribution of his former assistant, Alan Coulson, for his scientific contributions that underpinned the new project.[3] Moreover he agreed the new building could be named the 'Sanger Centre' on one condition: 'It had better be good.'[4]

Soon afterwards, a purpose-built larger centre was built on the same Hinxton Hall site. Through Sulston's initiative, backed by Aaron Klug, the Director of the Laboratory of Molecular Biology at the time, with financial support from Dai Rees of the Medical Research Council and Bridget Ogilvie of the Wellcome Trust, the future of the worm and human genome projects in the UK was assured. Although both worm and human genome projects became international collaborative efforts, there is no doubt that the UK took a lead in both these projects. It seems, in retrospect, that the Laboratory of Molecular Biology initiative in 1962 had been a model for what happened over 30 years later in 1993. Fred Sanger's scientific achievements, in the 1950s in working out how to sequence proteins, and in the 1970s in doing the same for DNA, were critical for the establishment of both laboratories.

A large number of automated sequencing machines were used, in both the public and privately financed initiatives, to sequence genomes. By this strategy the entire draft DNA sequence of the nematode worm *Caenorhabditis elegans* was finished by 1998,[5] and the draft human sequence was completed in 2000 and published in 2001.[6] Fred Sanger has been rightly called the 'grandfather' of the human sequencing project.[7] It was essentially his sequencing method that was used, even though he himself did not participate. However, two of Fred Sanger's long-term assistants, Alan Coulson and Bart Barrell, provided continuity after Fred Sanger's retirement in 1983 and made major contributions to the human and other genome projects. John, now Sir John, Sulston (Figure 66, right) was honoured by the award of a Nobel Prize in 2002, jointly with Sydney Brenner and Bob Horvitz, not for the human or worm sequencing project, but for earlier work establishing that cells die as part of the developmental process in the worm. There may never be a Nobel Prize awarded for sequencing the human genome. Although one of the most significant advances in biology ever made, it would be invidious to single out any one person or persons for this prestigious award. It was truly a collaborative effort involving many thousands of scientists worldwide.

The human genome draft sequence, announced in 2000, was hailed as one of the landmark discoveries in human biology and medicine, describing nearly 3 000 000 000 nucleotides sequenced at over 99.99% accuracy, even though some highly repetitive DNA was not sequenced. This was a huge accomplishment and a monumental effort by the teams of scientists involved. But important as it was, it was really only the beginning of a new era trying to understand the significance of so much information. Initially everyone wanted to know how many genes there were in a human being. Early estimates of 50 000 genes (protein and RNA coding) in humans were revised down to nearer 20 000 genes in

2001 – not too different from the initial estimate of 19 000 genes in the humble nematode worm in 1998. Since then these estimates have been revised upwards as techniques for studying the genome improve and data from other genomes, e.g. mouse, became available. Even so only about 1% of the human genome directly codes for proteins and stable RNA. However, there are recent data from the ENCODE project (an encyclopaedia of DNA elements), suggesting that at least 20% and possibly more than 80% of the human genome is involved in gene regulation in one way or another.[8] This concept that the 99% 'non-coding' part of the human genome might be 'regulatory' is a huge change in the way some scientists are thinking about the human genome. The ENCODE data are based on a number of different assays probing the whole genome (but excluding highly repetitive DNA). They include data on whether the sequence is copied into RNA, whether there are DNase I hypersensitivity sites and whether transcription factors bind to promoters or enhancer-like sequences in the genome.[9] The main concern with the new data is that they do not prove that such widespread RNA transcription in non-coding parts of the genome leads to gene regulation of any known genes. Moreover, most inherited genetic disease and cancers arise as mutations in the known 1% of the genome involved directly in coding for proteins. Thus the conclusion that most, or even that 20%, of the human genome is involved in gene regulation remains controversial. However, it is certainly possible that some non-coding DNA – a long way away from known genes – may be regulatory. We just need to know the target genes that are regulated. That research has yet to be done on a genome-wide basis.

Human beings differ from one another in sequence, even though the 7 or so billion people presently living on Earth now are all thought to be descendants of relatively few people who lived 100 000 or more years ago in Africa. How these early *Homo sapiens* evolved and gave rise to the variety of humans that now live on our planet is of great interest to evolutionary biologists. In particular it is of interest to know how variable the human sequence is between individuals. There are estimated to be about 4 million nucleotide differences (single nucleotide polymorphisms or SNPs) accounting for about 0.1% of the genome between a West African woman from Nigeria (a Yoruba) and a human reference sequence,[10] but still 99.9% of the genomes of any two people are identical. By extending sequencing to over 1000 individuals from diverse ethnic backgrounds, the total extent of nucleotide variation – including short insertions and deletions of nucleotides – is estimated to be nearly 40 million nucleotides.[11] Thus more than 1% of the human genome can vary in sequence. Even this 1% figure is probably an underestimate, since the extent of variation would be expected to rise as more human genomes are sequenced.

New gene sequencing strategies

Even before the draft human genome was announced in 2000 there were initiatives that envisaged the need for sequencing strategies that were more efficient and more cost-effective than the Human Genome Project. This is estimated to have cost many hundreds of millions of pounds of public and private finance since it was announced in 2000. The aim now is to re-sequence the human genome for less than $1000. This would enable sequencing to be part of routine medical practice. We might envisage that in ten years' time each one of us might hold our own genome sequence on our own personal smart phone, with apps to interrogate our genome when we visit our doctor for advice about medical issues. Did we have genetic variation that predisposed us to such defects as heart attacks, type II diabetes or cancer, for example?

At present, there are a number of so-called 'massively parallel' sequencing procedures available. In 2012 costs were about $8000,[12] though they are reducing all the time. The goal of sequencing the human genome for $1000 is rapidly approaching and is likely to be achieved within the next year or two. The basic idea of massively parallel sequencing is to sequence many millions, or even billions, of DNA molecules simultaneously. This concept differs from the procedures that were used in the Human Genome Project where mixtures of DNA molecules were usually purified by 'cloning' in living organisms, such as bacteria or yeast, and then individually sequenced from the separated 'clones'. Thus new-generation sequencing procedures can sequence short stretches (say) 100 nucleotides of many millions of individual DNA fragments from the human genome in 24 hours at high accuracy.

Massively parallel sequencing requires two other criteria to be fulfilled. Individual DNA molecules must first be immobilised in some way on a two-dimensional surface of some sort so that a sequencing procedure can be performed on millions of such DNA molecules at the same time. Second, a sequencing reaction must be performed on the whole set of immobilised DNA that distinguishes which nucleotide is added at each cycle of DNA synthesis. This is determined by automatically monitoring which of the four nucleotides – C, A, G or T – is added to each of the millions of DNAs that are being sequenced. Most, but not all, sequencing methods involve detection of fluorescence linked to the addition of one nucleotide at a time in a primed sequencing reaction. By using fluorescent nucleotides with chemically different fluorescent properties (i.e. different emission spectra), which are excited by lasers, one can distinguish which of the four nucleotides is involved in a given cycle.

It is usual, although not obligatory, in all the new sequencing strategies to amplify individual DNA molecules into a 'library' of the DNA of interest using polymerase chain reaction (commonly abbreviated to PCR), a technique invented by Mullis in 1983 for which he obtained the Nobel Prize in Chemistry in 1993.[13] The polymerase chain reaction technique is a way of amplifying double-stranded DNA by the use of two DNA primers, a forward and reverse primer (usually between 10 and 30 nucleotides in length) and a DNA polymerase, to obtain many millions of copies of an original DNA. Because DNA is (usually) double-stranded it must be heated between amplification cycles to denature it in order to separate the two constituent single-stranded DNAs. Originally PCR was very inefficient because the DNA polymerase did not survive the heating steps. This DNA polymerase enzyme had to be reintroduced in each step of the 30 or so cycles to complete the amplification making the procedure cumbersome and time-consuming. Many improvements in the method have been introduced since. The most important innovation was the introduction of a heat-stable DNA polymerase, Taq polymerase, isolated from a bacterium, *Thermus aquaticus*, found in hot springs. This polymerase survived the repeated heating steps and it was only necessary to add the enzyme once at the beginning of the procedure, instead of at each amplification step. A second important innovation was the use of more accurate DNA polymerases, e.g. 'Phusion' DNA polymerase, thus minimising errors made by the polymerase in copying the DNA sequence. The first heat-stable polymerase, Taq polymerase, was prone to such errors. Although errors are still made in copying by the polymerases used for PCR, the rate of error is extremely low, if low-error prone polymerases are chosen.

An important development of PCR was *emulsion PCR*. This technique was introduced in 2003 by Vogelstein's group as a way of compartmentalising single molecules of DNA templates, while at the same time allowing these single molecules to be massively amplified on magnetic beads.[14] Thus there were many copies of individual DNA molecules attached to beads, so that there was enough DNA to sequence in ordered arrays. Several sequencing methods have used emulsion PCR to amplify DNA prior to sequencing.

Another significant development was the ability to efficiently attach oligonucleotides to glass.[15] This method allowed amplification to be carried out on the glass surface and this technology is used in the 'Illumina' sequencing method.

454 or *pyrophosphate sequencing* was the first widely available, massively parallel sequencing method to be developed. It uses naturally occurring triphosphates, without terminator or fluorescent modification. The addition of a single triphosphate, either C, A, G or T, to a polymerase extension of a primed amplified template, attached to beads arrayed on a plate, is detected in real time by the analysis of the pyrophosphate released as the nucleotide is polymerised.[16] However, 454 sequencing has been largely superseded by the *Illumina* technology.

Illumina uses technology acquired in 2007 and has proved to be the most popular high-throughput sequencing method over the last few years. The basic technology was developed by Solexa, a Cambridge-based company set up by Shankar Balasubramanian and David Klenerman of the Department of Chemistry at the University of Cambridge. Unlike the dideoxy technology invented by Fred Sanger in 1977, Solexa developed the concept of a *reversible* nucleoside triphosphate terminator with an *O*-azidomethyl group blocking the 3′ position of the sugar residue as well as a *reversible* fluorescent (FLUOR) dye attached to the base moiety of the triphosphate (Figure 72).[17] A different fluorescent dye is used for each of the four triphosphates so that a camera can monitor which of the four dyes is added in a primed synthesis reaction with DNA polymerase. The basic sequencing idea is still similar to the Sanger procedure of 1977, in which a nucleoside triphosphate is incorporated into a growing DNA chain by a copying reaction catalysed by DNA

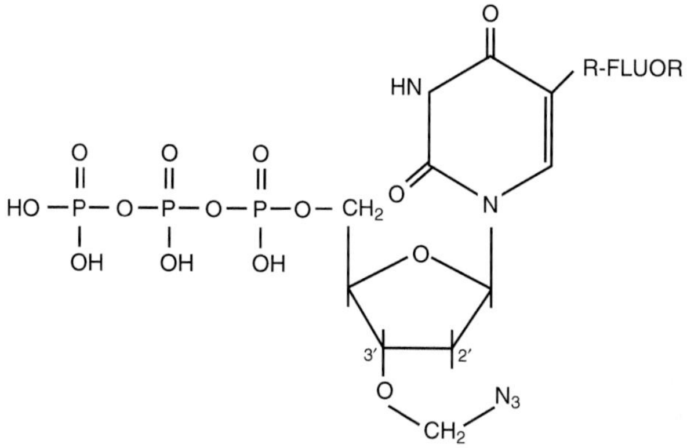

3′ O-azidomethyl dTTP

Figure 72. Reversible terminator modified dTTP with FLUOR attached to the base. (Redrawn and modified from Bentley DR *et al.* (2008) *Nature* **456**: 53–59.)

polymerase (Figure 56). But there is a difference. Because of the presence of the reversible terminator, it is possible to simply mix the four different triphosphates (i.e. CTP, ATP, GTP and TTP), modified to contain the appropriate terminators, and then monitor by fluorescence which of the four triphosphates is added in a *single* step of polymerisation with DNA polymerase. Although sequence reads are relatively short (100 nucleotides) the method can rapidly generate accurate sequence information equivalent to the length of the human genome in a day or so.

In outline the *Illumina* method, as explained to me by Vice-President, David Bentley (Figure 73), involves the attachment of a library of DNA, along with suitable primers to a glass slide such that the DNA is amplified to form DNA clusters. The slide would typically contain many millions of these clusters. To determine the sequence, a mixture of the four terminator triphosphates – CTP, ATP, GTP and TTP – are added and DNA polymerisation allowed to occur. Then, unincorporated triphosphates are washed away. A digital camera takes images of the fluorescently labelled DNA. Then the terminal 3′ blocking group on the sugar along with the FLUOR on the base is chemically removed from the DNA, allowing for the next cycle to begin.

Another recently described method is *ion torrent* sequencing.[18] This method dispenses entirely with fluorescence and uses unmodified naturally occurring triphosphates without fluorescence or chain

Figure 73. David Bentley at Illumina, Cambridge, 2013.

terminators. It is the first method to be fully digitised. If a nucleotide is incorporated by the DNA polymerase a hydrogen ion (H^+) is released causing a drop of 0.02 pH units. This pH drop is detected and transduced by a semiconductor into a digital signal, recording the sequence directly on a computer.

Other ingenious methods have been developed by *Pacific Biosciences*, who have been able to sequence single molecules of DNA, without PCR amplification, in tiny holes holding minute volumes (10^{-21} litres) of liquid containing a DNA polymerase to which the DNA to be sequenced and a primer are bound.[19] *Life Technologies* and *Complete Genomics*[20] have developed a clever method based on sequencing *by* ligation. These methods are less widely used now.

Finally, *nanopore* sequencing[21] – still under commercial development by Oxford Nanopore in 2014 – uses an entirely different principle of sequencing. This method is very simple in concept and records which of the four nucleotides is present at a given position in a single-stranded DNA as it is electrophoresed through a pore embedded in a membrane. The pore is formed by a transmembrane protein that can distinguish C, A, G and T in DNA, due to interactions between the four different nucleotides and amino acids in the pore causing a small change in current. Modified bases, such as 5-methyl cytosine, are also detected.

Obviously, a critical issue in all sequencing methodologies is sequence accuracy. What is the rate of error? All high-throughput methods have a lower accuracy than the traditional dideoxy method used in the original human and worm sequencing projects. In these original projects accuracy was estimated as greater than 99.99%, or less than 1 error in 10 000 bases. However, in high-throughput methods one can compensate for their intrinsic lower accuracy by the strategy of 'high coverage'. This term is used to indicate that the genome is sequenced many times over. Thus a relatively high error rate of say 0.1% (1 error, on average, every 1000 bases) in sequencing 100 nucleotides may be acceptable, if enough fragments are sequenced to cover the genome 10 times or more, thereby reducing the error – assuming the errors are random – to 0.01%.

In practice, if the error rate is 0.1%, it is necessary to sequence the human genome more than 30 times over to ensure a minimum of 10 times coverage over virtually the whole genome. This 30-fold coverage is necessary because of sampling errors and other potential biases in the methodology. Even then some bases may be missed or misassigned because of repeat sequences (duplicated genes or pseudogenes[22]). Whether an error rate of 99.99% accuracy is tolerable may also depend on whether the application is *de novo* sequencing or one is re-sequencing (say) the human genome looking for variation or disease-causing mutations. Of course this error rate may not be uniform over the whole genome. The methods for generating DNA to be sequenced, however, have become much less biased than they were when clones were isolated from living organisms. By physically shearing DNA and then 'repairing' the DNA ends suitable for the ligation of 'adapter' or 'linker' molecules, it is likely that such 'libraries' of DNAs are a reasonably random representation of the genome to be sequenced.

Further details of some of the new generation sequencing methods and technology related to these methods is presented in specialist reviews.[23]

8

Cancer: the impact of new-generation sequencing

If we know the complete sequence of the human genome and understand how it works, and how tumours are formed, then we can begin to understand how to deal with tumours better.[1]

Both of Fred Sanger's parents died of cancer at relatively young ages, his father, Frederick, aged 60 in 1937 from an operation to remove a stomach cancer. His mother, Cicely, aged 58, died a year later from cancer of the colon. Fred Sanger was a young man of 20 then – still studying at Cambridge for his first degree. At that young age, he could not have imagined that later in his life it would be his research that would make the dream of sequencing the human genome a reality. In 1938 it would have been a leap of faith to predict that by the beginning of the twenty-first century, the entire human genome would have been sequenced.

Fred rarely allowed himself the luxury of predicting future results of research. 'I don't want to look into the future very much. I'm a scientist not an astrologer! I don't like to predict things you can't be too sure about.'[2] He believed in evidence, not in theories of what might be possible and how long it might take to solve a problem. But we know that Fred hoped that his research might lead to medical advances even though he rarely spoke about this in public. In 1992, in an interview with a journalist, he said: 'If we know the complete sequence of the human genome and understand how it works, and how tumours are formed, then we can begin to understand how to deal with tumours better.'[3] The knowledge that cancer had caused the early death of both his parents may have been in the back of Fred's mind when he made this remark.

In the last chapter I described the new-generation DNA sequencing methods introduced since Sanger developed his dideoxy sequencing method in 1977. I have chosen, as an example, to outline current knowledge from these new-generation methods in one type of cancer – breast cancer. This information is novel and challenging. It is giving us unprecedented detail at the molecular level of the genes mutated in breast cancer. It is a medically important application of DNA sequencing, which is being applied not only to breast cancer but to a whole range of human cancers, which I also mention briefly. Finally I outline a remarkable result and a new specific treatment for a particular mutation in melanomas – a type of skin cancer caused by exposure to sunlight. This therapy was developed from knowledge gained by sequencing the DNA of melanoma patients.

Historical perspectives of breast cancer

Breast cancer is the commonest cancer in women and has been studied and treated for many years. These cancers are uncontrolled abnormal growths of the breast tissue, usually of the cells lining the ducts of the breast. They can be benign or malignant. Benign tumours do not spread to other parts of the body, whereas malignant tumours can spread. Most breast cancers are *somatic* mutations. In other words they develop during the *lifetime* of the woman who has the cancer. Somatic mutation as the major cause of cancer was suspected nearly 50 years ago,[4] although proof of the causal nature of somatic mutation in cancer had to await the development of DNA sequencing. But a small proportion of women have a *genetic* mutation inherited from one of their parents that causes breast cancer. This is often described as *familial* breast cancer since it runs in families. Unfortunately, in the case of familial breast cancer, women have a high risk of developing breast or ovarian cancer in their lifetime.

There were indications, nearly 150 years ago, that breast cancer was inherited in some families. In 1866 Paul Broca recorded that ten of the 24 women of five generations of his own family in France died of breast cancer.[5] Later, in 1926, Janet Lane-Claypon (arguably the first epidemiologist), working in the UK in a survey of 500 cancer patients and 500 controls, concluded that there was an inherited risk of breast cancer in some families.[6] Nevertheless these findings, although reproduced by other studies early on, were not generally believed by clinicians. It was possible that the increased susceptibility to cancer in some families, apparently caused by a familial risk, was simply a coincidence, since breast cancer was a common cause of death. Another possibility was that breast cancer was caused by environmental rather than inherited factors. For example, in the 1940s, Bittner had shown that mice infected with a mouse mammary tumour virus could acquire breast cancer from infected milk.[7] At that time it was thought possible that a human virus, related to the virus causing the tumours in mice, might have caused breast cancer. As it turned out, this viral hypothesis of breast cancer was incorrect.

That breast cancer could be inherited was not finally proven until the mid-1990s when defects in two genes, breast cancer type 1 and 2 susceptibility protein genes, abbreviated *BRCA1*[8] and *BRCA2*,[9] were identified. The functions of the proteins expressed from these genes, although quite distinct, are both involved in the repair, or correction, of double-stranded breaks in DNA. Normally, DNA is copied into progeny DNA by an enzyme in the process of cell division without any errors, but occasionally errors or DNA breaks occur during the copying process. These two genes are responsible for repairing these mistakes. Failure to correct these errors, and the resultant accumulation of errors in many genes, is believed to lead to the increased risk of breast and other, particularly ovarian, cancers.

Defective *BRCA1* and *BRCA2* genes are now well-recognised inherited risk factors for breast cancer. However, there are other even rarer, *inherited* defective genes that increase the risk of women developing breast cancer in their lifetime. One of these is the gene coding for a tumour protein, abbreviated *TP53*.[10]

Sequencing of cancers: identification of genes commonly associated with breast cancer

New-generation sequencing is a powerful new method for finding out the actual defects at the gene level causing the cancer. Clinicians have long recognised that breast cancer is a variable disease, with variable outcomes in patients, but the reasons for this variability have been imperfectly understood at the molecular level. Now it is straightforward, as long as biopsy of the cancer is available, to sequence the complete genome of the cancer cells to find out the mutations in their DNA. The basic idea is to sequence the entire DNA, or at least all the coding regions of the DNA, isolated from the cancer tissue of the patient. The question can be put another way. Which of the 20 000 or so genes in the human genome is faulty?

If there were one gene mutated in breast cancer, molecular classification would be quite simple. The remarkable result is that many different mutations, in many different genes, are found in breast cancer. Breast cancer turns out to be a genetically extremely diverse disease. For example in a number of studies of breast cancers published in 2012,[11] defects were found in many different genes clearly demonstrating that there are multiple causes of breast cancer at the genetic level, although the most frequently mutated genes were common to most surveys. This suggests that there may be a spectrum of causes with some commonly mutated genes but many more genes that contribute to the cancer. Most breast cancers differed from all others and had multiple affected genes but a few had only one or two affected. This does not mean that there are an infinite number of genes causing breast cancer, because more than 50% of all patients in one study had defects in at least one of seven genes.[12] From our understanding of the function of these seven genes, it is clear that these gene defects could influence the mammary gland and cause uncontrolled growth leading to breast cancer. But this interpretation is likely to be only part of the story and is certainly oversimplistic, because defects in many genes – besides the seven genes mentioned above – contribute to breast cancer.

At the time of diagnosis of breast cancer in a patient, it is likely that the tumours will already be a mixture of different cancer cells with different properties. In other words, the 'founder' or 'driver' causative mutation may well be masked by further 'passenger' mutations arising as a result of secondary events, either further independent somatic mutations or mutations arising as a result of the first mutation. Thus the breast cancer sequence may not be derived from the DNA of a single cancer 'clone', but rather it may derive from mixtures of 'clones' in varying proportions with different mutations. What is unclear at present is the relative importance of the different mutations to the cancer occurring in patients. Even more importantly, we do not yet know which gene mutation or mutations are critical for the spread of breast cancer in the body causing the primary tumour to become malignant.

It should also be emphasised that many different types of mutations, even within a single 'driver' gene, can lead to cancer. They may be point mutants affecting a single nucleotide. They may be relatively short insertions or deletions of nucleotides, generally from about one to 20 nucleotides in length. They may be whole-gene amplifications or whole-gene deletions. Mutations may exert their effect in different ways. Mutations may interfere with the synthesis of messenger RNA from DNA. They may interfere with mRNA processing, in particular splicing out of introns – regions of DNA that interrupt the coding of proteins and have to be removed or spliced out in precursor mRNA. Finally, the effect of mutations may be to interfere with the function of the protein so that the protein is completely inactive, but in other cases some partial function may be retained. The precise nature of the mutation – even within a single gene – may have different consequences causing a more benign or a more aggressive cancer.

Do *both* copies of cancer genes have to be defective to cause cancer? It is thought that for most cancer genes there would have to be a somatic mutation in both the paternally and maternally inherited copy of a gene for cancer to develop. In the case of women who develop breast cancer from defective, inherited *BRCA1* and *BRCA2* genes, the situation is different because here only one copy of the defective gene is inherited. The other copy is normal. Breast cancer arises because the other, normal, copy becomes damaged or mutated in the breast tissue. Amplification of genes is another quite common defect in breast cancer.

Pan-cancer sequencing

Given that breast cancer is itself a highly variable disease, with multiple affected genes, it is likely that there would be even greater numbers of causative genes involved in a whole range of different cancers. Evidence for this comes from a survey of 12 different types of common cancers in 2013.[13] In this study the authors analysed more than 3000 tumours across 12 tumour types, including breast, ovarian, lung, colon and kidney cancers. Overall they identified over a hundred significantly mutated genes. They, and many other authors, now emphasise that there are many examples of defects in genes that are typically associated with the maintainance of cell structure and metabolism, e.g. genes involved in chromatin modification, or in RNA splicing or in proteolysis. Thus many cellular and enzymatic processes in the metabolism of the cell, besides transcription factors and cell signalling, are involved in tumour formation.

It is interesting, but unsurprising, that some affected genes were common to many types of cancers. The transcription factor *TP53* gene was commonly mutated in at least 40% of all 12 tumour types, and another gene involved in signalling, *PIK3CA*, was mutated in about 20% of all tumours. Colon cancer, from which Cicely Sanger, Fred's mother, died, is generally linked to defects in signalling genes as well as *TP53*. Stomach cancer, from which Frederick Sanger senior died, could have been linked to infection with a microorganism, *Helicobacter pylori*, that is known to predispose to stomach ulcers and sometimes to cancer.

Melanomas and their therapy: a sequencing approach

Melanomas are cancers of melanocytes, cells in the skin and less commonly elsewhere, that synthesise melanin – the black pigment that defines the colour of our skin. In black people melanin protects the skin from the harmful effect of sunlight and they rarely get melanomas. Light-coloured people, particularly northern Europeans, are susceptible to skin melanomas caused by UV damage to DNA of melanocytes, or melanocyte stem cells, through exposure to sunlight. Melanomas are highly malignant and spread rapidly. Therapy is difficult and prognosis poor unless surgical intervention is performed early on.

In 2002 DNA sequencing of candidate genes[14] showed that about half of malignant melanomas had a particular mutation in the DNA of the *B-RAF* gene causing an amino acid change at position 600. B-Raf is a protein involved in intracellular signalling of melanocytes, and many other cells in the body. This changed a valine (V) residue in B-Raf to an acidic, glutamic acid (E) residue, abbreviated V600E. B-Raf, instead of being downregulated in the signalling cascade, was now permanently active and functioned as a kinase,[15] phosphorylating the next protein in the signalling cascade, thus overriding the normal controls that regulate its activity. Thus the melanocytes divided out of control and proliferated to cause the cancer.

Remarkably, an inhibitor was discovered in 2008 that specifically targeted the mutant B-Raf protein by locking its 'kinase domain' in a configuration that was not observed with the normal, wild-type B-Raf protein. The authors described this new drug as the first (small-molecule) therapeutic molecule selectively inhibiting a target that exists only in tumours.[16] Clinical trials have confirmed that this kinase inhibitor can prolong life for such patients.[17] Although side reactions and resistance to this inhibitor have been observed, this therapeutic approach will hopefully pave the way for the development of other drugs specifically directed to cancers.

Conclusion

Scientists can now sequence the cancer genome to find out the actual genetic defects in cancers. This information is likely to be taken into account to determine the best possible treatment for patients in the future as sequencing costs decrease and are applied in the clinic. But much more needs to be done to target the specific genetic defects, or the consequence of these defects, in the various types of cancer so as to devise drugs that will specifically target the defective protein, or correct an over-abundance or lack of a functional protein, without harming the normal unaffected non-cancerous cells of the body. This is a difficult, but not impossible, task as the specific treatment for myeloma has shown. Real, directed research in the treatment of cancer is only just beginning. Humanity has to be indebted to Fred Sanger whose research made this possible.

9

Commentaries on Fred Sanger's scientific legacy

At the suggestion of Cambridge University Press – after my manuscript was completed – I commissioned five short commentaries by distinguished molecular biologists who had read my biography. These complement the Foreword by Sir Edwin Southern.

The commentators are:

Paul Berg, *Cahill Professor of Biochemistry, Emeritus, Stanford University, Stanford, California, who shared the 1980 Nobel Prize with Fred Sanger and Walter (Wally) Gilbert.*

Elizabeth (Liz) Blackburn, *Professor of Biochemistry and Biophysics, University of California, San Francisco, who studied for her PhD with Fred Sanger in Cambridge and is a Nobel Laureate (2009).*

Sir John Sulston, *founding Director of the Wellcome Trust Sanger Institute, Hinxton, Cambridge, UK, who worked on the nematode and human genome projects. He is a Nobel Laureate (2002).*

David Bentley, *Vice-President and Chief Scientist, Illumina Inc., Chesterford Research Park, Essex, UK, who gives an industrial perspective. He worked on the Human Genome Project with Sir John Sulston before gaining industrial experience with Illumina, located near Cambridge, UK.*

Sir Paul Nurse, *currently President of the Royal Society, London and a Nobel Laureate (2001).*

Paul Berg

George Brownlee's intimate biography of Fred Sanger emerged from their shared scientific collaborations and a series of wide-ranging interviews conducted long after Sanger had retired from active research. There is the sense of two old friends sharing moments in history. The interviews reveal a unique personality, often self-deprecating but undeniably confident of his creative powers and proud of his accomplishments. Sanger's early schooling, undergraduate and graduate experiences at Cambridge reveal little of the daring and ingenuity that were to characterise his remarkable scientific achievements. His PhD thesis, carried on during the war's early years, was by any measure rather undistinguished but his subsequent accomplishments as a member of the Cambridge biochemistry faculty were astonishing and explosive in their impact. Beginning with the limited goal of identifying the terminal amino acids of the insulin molecule, Sanger developed methods for that purpose that ultimately led to solving the protein's complete amino acid sequence and its non-peptide linkages. That accomplishment, arguably, initiated the modern field of protein chemistry and focused attention on how protein chains fold into specific three-dimensional structures compatible with their functions. His findings also refocused attention on the necessity for a universal mechanism for synthesising the myriad cellular proteins, each with a defined order of amino acids.

The discovery that proteins are assembled under the guidance of genes and that that process relies on nucleic acids, DNA and RNA, set Sanger's eyes on sequencing those molecules. Having received the 1958 Nobel Prize in Chemistry for the insulin work at the rather early age of 40, he recognised that he now had the time and resources to reach for that goal. 'Hearing' Sanger describe how he conceived of the approaches he explored and how he adapted or invented new tools that enabled him to sequence a small tRNA and then longer viral RNA genomes is dazzling and inspiring. Every student of molecular biology would profit greatly by lingering over Sanger's own descriptions of the several strategies he attempted and the tools he had to invent to arrive at what is referred today as Sanger DNA sequencing. For that achievement, Sanger was awarded a second Nobel Prize in Chemistry in 1980, one of very few people ever with that double distinction.

His success transformed molecular biology in inestimable ways, for today most biological investigations where genetic analysis is the underlying focus rely on DNA sequencing as the primary experimental paradigm. Moreover that methodology is equally foundational in such varied fields as evolutionary biology, plant breeding, palaeontology, anthropology, linguistics, forensics and computer science. Sanger sequencing was at the core of the multinational effort to sequence the human genome and the genomes of innumerable other organisms and even of ancient humanoid and prehistoric species. Further, DNA sequencing is at the core of modern bacteriology and viral epidemiology. Even more extraordinary is the extent to which Sanger DNA sequencing technology is transforming medicine broadly but more specifically in cancer biology by facilitating the detection of DNA sequence alterations that signal predispositions to cancer, by identifying sequence changes that confirm overt cancers and monitor responses to therapy. Fred Sanger was motivated by the challenge of discovery and believed fervently that through art and science, in their broadest senses, it is possible to make a lasting contribution towards the improvement and enrichment of human life. He cherished his good fortune in having been a major player in that enterprise.

Elizabeth Blackburn

The term 'hands-on' scientist is widely used. It is hard to think of anyone in the modern biological research era who so exemplifies the value of this kind of scientist, and yet is so little known to the world at large, as Fred Sanger.

Frederick Sanger (1918–2013) was a rarity in many ways. He was awarded the Nobel Prize in Chemistry not just once, but twice, for pioneering the sequencing of the fundamental chemical building blocks that make up proteins and the genetic material, DNA. Fred's first Nobel Prize (1958) was for determining, for the first time ever, the complete sequence of the chemical building blocks (amino acids) in a protein (the hormone insulin). Fred then focused his interest on sequencing the building blocks (called nucleotides) of nucleic acids: DNA as well as its cellular information carrier RNA. His second Nobel Prize in 1980 recognised his pioneering methods for sequencing DNA.

Fred was also unusual in having not only a nose for central questions in biology, but also an uncanny degree of acumen for chemistry and the ability to focus this on direct experimental approaches to such questions. The macromolecular proteins and DNA are absolutely central to life on planet Earth; proteins carry out the chemical reactions of life and constitute much of the material of cells of our bodies, and of all cells in living creatures. The material form of the information carried by the genetic material is, quite simply, the sequence of nucleotides in DNA, which encompasses codes of various kinds, most notably the code that determines the amino acid sequences of proteins. That code had been brilliantly but only indirectly deduced by others when Fred started his work on DNA sequencing methods. Fred presciently saw that direct sequencing was the key to understanding how DNA does its work.

As George Brownlee's succinct and wonderfully illuminating biography of Fred Sanger describes, Fred's contributions were deceptively simple but profoundly far-reaching. Even now it is hard to recall the time when it was not known if proteins had specific sequences of amino acid building blocks. Yet such was the situation when Fred set out to determine the sequence of insulin in the 1940s. He had to invent methods to do this. Fred sequenced insulin by direct chemical analyses, using ingenious chemically based laboratory methods that he adapted or devised. These were in and of themselves significant contributions, but the conceptual advance was also immense. By sequencing insulin, Fred provided proof that a protein was a defined chemical entity made up of a set sequence of amino acids joined together into a long linear chain (or, as it happened in the case of insulin, two chains held together by special chemical bonds of a type now known to occur often in proteins). His work laid the ground for a concept that underpins modern biology: for each of the myriad proteins that exist, it is its own specific amino acid sequence that determines how that protein will assume its special shape, thus allowing it to function.

It is also hard to recall how recently it was that there was no known way to sequence a genome. The genome is the term used to describe the entirety of DNA that makes up the genetic blueprint of an individual, be it a person or any living organism. Although the genomic DNA is usually divided up in the form of chromosomes, still the DNA of each chromosome typically is extraordinarily long. Thus Fred had to find ways to tackle this seemingly overwhelming technical problem. He initially explored it from different angles. As George Brownlee's biography of Fred Sanger describes, in the early 1970s Fred had assembled a team of researchers in his group who successfully developed a variety of techniques for sequencing RNA molecules (whose nucleotide sequences are copies of those in DNA), portions of the relatively small genomes of RNA viruses, and small stretches of DNA. But in the end, as George Brownlee's book describes, it was another unusual attribute of Fred, his hands-on approach, that was the key. Fred himself, working every day at his bench on developing his DNA sequencing technique (assisted by just one or two technicians), brilliantly solved the problem. His solution to sequencing DNA, which he had devised by the mid-1970s, was the one that that ushered in the current era of genomic sequencing. This is because his approach could become adapted as the basis of the high-throughput, automated DNA sequencing methods needed for sequencing the enormous numbers of building blocks in each genome of people, and most living organisms.

In addition to highlighting Fred's scientific accomplishments, George Brownlee's book also paints the first real picture the world has seen of Fred as a person. I think that these two facets of Fred were inextricably linked. I had the great good fortune to be one of Fred's Cambridge University PhD research students. I was in his group at the Laboratory of Molecular Biology in Cambridge during the years of 1971 to 1974, when Fred was himself working out what became his widely used and adapted DNA sequencing methods. He had an enormous influence on my future scientific life in at least three ways. Firstly, Fred's hands-on attitude to science was immensely inspiring and admirable to me. Secondly, as George Brownlee's biography clearly brings out, Fred was an understated person who was deeply rooted in moral values of truth and humanity. Yet he rarely if ever explicitly stated these values and certainly did not preach about them. Rather, he simply chose to live these values and convey them to others round him through his own practical example, as lived out though his science. This was particularly appreciated by those of us who knew and worked with him in his professional life, and I am surely not the only one who has tried to emulate his example.

Thirdly, my experience in Fred's lab proved invaluable for my own subsequent research on telomeric DNA and telomerase. Fred's lab was probably one of the best places to learn and be exposed not only to his ultimately used technique for sequencing DNA, but also to his earlier methods for sequencing nucleic acids, and to the other approaches that were being concurrently explored and exploited by others in his research group. En route to his success in finding out how best to sequence DNA, Fred had assembled a lively and international lab group. It was their multiple approaches to piecing together the (relatively short) RNA and DNA nucleotide sequences that provided me with a toolbox to sequence telomeric DNA. This amalgam of different methods was needed to sequence telomeric DNA directly, because of the structural peculiarities of DNA at the very ends of chromosomes. I did this for the first time as a postdoctoral fellow in Joe Gall's lab at Yale University. This in turn allowed the discoveries of the generality of telomeric DNA and of the novel enzyme telomerase (with Jack Szostak and Carol Greider, respectively, for which the three of us shared the 2009 Nobel Prize in Physiology or Medicine). For example, when in my own lab my student Carol Greider and I were hunting for a then-hypothetical enzymatic activity (which we later called telomerase) that could synthesise telomeric DNA, the patchwork of methods devised by Fred's lab members to piece together short stretches of DNA sequences again rose to the fore and allowed us to confirm the sequence of products of this enzymatic activity. This research has turned out to have wide implications in biology. The maintenance of telomeric DNA is essential for eukaryotic cells. Furthermore, its impairment has now emerged as a common mechanism that underlies and interacts with other etiologies of human chronic diseases of ageing. In this way, Fred's influence extended even further, not only from his own commonly used DNA sequencing methods, but also from the fertile ground of his research group.

I greatly liked Fred's style of being a PhD advisor: he was there at the bench in his own small lab almost every day, quietly working. He would readily talk about their experimental work to his students or any one in his group who asked, but not for too long, as he was happiest getting on with his own bench work. His interactions with his research group reflected his implicit message that he trusted his trainees to succeed, and on their own terms. He interacted kindly with people and the only signs of mild impatience I saw were if he thought someone was being pretentious or unintelligent. Yet although Fred was quiet and low-key, he had a sly sense of humour and sometimes would let out a dry comment. I recall that at some point, in this embryonic stage in nucleic acid sequencing, Fred did not think very highly of overly elaborate instrumentation. On one occasion a handsome large new centrifuge was installed just outside his lab area. I and a couple of his lab group members were inspecting it; Fred walked by carrying one of his experimental samples, and when we drew his attention to the centrifuge, he glanced at it and said: 'Too many knobs.' I peered at it; it had three. By his own example of the way he did his experiments, he conveyed his unstated philosophy: he believed that the best approaches were directly experimental and hands-on, with as little as possible separating the experimentalist from the experimental material. However, he was no Luddite and appreciated what was needed to get a job done: when some years later the Sanger Centre was opened – the epitome of the elaborate instrumentation needed to accomplish high-throughput genomic sequencing – he was very gratified and intrigued by it.

Fred Sanger was truly an extraordinary and wonderful individual, and his influence in modern biological and biomedical sciences cannot be overstated. As time goes on, I predict that its importance and range will become ever more appreciated.

John Sulston

If you type the word 'Sanger' into a search engine, you may find that the Sanger Institute, Hinxton, Cambridge – named after Fred – comes out on top. I think he wouldn't have minded that. He has a living memorial – better than a statue, really – where work that he started goes on, as it does in his old laboratory, the Medical Research Council Laboratory of Molecular Biology, Cambridge and all over the world, wherever people delve into the complex mechanisms of life. There is no need to ask what specifically has come out of Fred's sequencing, because the answer is too broad. If you ask a teacher 'Why are you showing that child how to read?' the reply might be 'To open the world's knowledge, so that the child can understand and achieve more.' Fred taught us to read the information of life, so that we can begin to understand it.

In this lovely biography, George Brownlee brings out the essence of Fred and shows us what made him so special – his love of science, of doing, of steadily persisting at important things. In Fred's own autobiography,[1] he writes: 'I consider that my own personal contributions have been to the methods. In the course of the work we have determined many sequences and obtained significant results, which have been reported elsewhere, but in these I have usually been part of a team.' Then, considering his retirement: 'I think that if I had gone on working I would have found it frustrating and have felt guilty at occupying space that could have been available to a younger person.' For an age in which research is often judged on market value, and in which working to the end is regarded as a badge of honour, Fred had wise perspectives. He reminded us that there is more than one way to live well and do great science.

David Bentley

Fred Sanger's innovation and achievements enabled us to read the code of life. His breakthroughs transformed science, medicine, industry and society, and continue to open our eyes to a new knowledge and understanding of how life works.

The DNA and RNA sequence lies at the very heart of life. It is the nature of genetic information, and is used in every cell in every living thing. The beauty of this code is the simplicity with which the information is stored. The four-letter alphabet of the DNA bases is used to spell out every biological message, often by translating the information to specify the order of amino acids in proteins. Fred Sanger gave the world a way to read this information, not once, but twice. He tackled the proteins directly first, working out how to determine the order of the amino acids that form insulin. He then switched to the study of DNA, which provides the underlying code for insulin and every other protein. Fred's innovative skills in chemistry led to a method to decode DNA and RNA. Fred delved into the natural process of how the DNA chain is copied in nature, and adapted it. He manufactured new DNA building blocks that were lacking an important oxygen atom on the third carbon of the ribose sugar ring. This caused the process that copied the DNA chain to terminate whenever it incorporated one of Fred's building blocks (or 'chain terminators'). The termination was permanent, and the complex mixture of terminated chains was then separated out in order to read off the DNA sequence. In retrospect, this was Fred's biggest breakthrough.

Biology and human health are closely intertwined. Understanding how life works provides a baseline for studying what happens when something goes wrong. A protein such as a blood-clotting factor changes, becomes ineffective, and causes haemophilia. A pathogen such as influenza virus changes, escapes immunity, multiplies and causes infectious disease. A normal human cell suddenly changes and grows uncontrollably as a cancer. These changes happen first at the molecular level. The DNA or RNA mutates: one or more of the bases is altered or lost. Fred's sequencing method enabled us to identify all such disease-causing mutations. It revolutionised our ability to understand and diagnose disease, and laid the foundation for a new era of genetic medicine. Sequencing has also driven progress by enabling scientists to determine the entire genetic make-up, or genome, of many living things. Most notable of these is the genome sequence of *Homo sapiens*: a near-complete, public catalogue of all the genes and other sequences in our chromosomes – a catalogue that underpins genetics today.

The birth of industry, computing and the internet are well-known examples of new technologies that have changed society. DNA sequencing can now be added to this list. The widespread occurrence and utility of sequence information in nature brought an essentially unlimited demand for sequencing. This demand soon outstripped the capacity that was available, even after the highly successful industrialisation of Sanger's method in the 1990s. At the turn of the twenty-first century, there was a further revolutionary advance in the field of DNA sequencing. Another group of Cambridge chemists developed a method to carry out DNA sequencing that used the concept of terminators, but made them reversible so that when each DNA chain was copied, the process could be stopped and started at every base. Each terminator blocked the copying process, allowing the base to be read, and the termination was then reversed to allow the process to be repeated. The insight offered by Fred's sequencing method inspired another era of innovation that has enabled scientists worldwide to sequence a billion DNA molecules on a single microscope slide in a few days. The explosive growth of this technology (over 1 million-fold in eight years) has created a new global industry in DNA sequencing. Sequencing technology is laying the foundations for precision medicine. Individual genome sequencing is becoming affordable and accessible as a new tool in research and healthcare. Diagnosis and treatment of diseases will increasingly benefit from the ability to read a patient's underlying genetic make-up, and sequencing will also enable us to monitor other living things in our environment, from the elephant to the microbe. Sequencing tells us about life, and will give us a new insight into the world we live in.

This book provides a rare opportunity to see a glimpse of a remarkable man who changed the world. It reveals his desire to build things, to pursue the truth, to listen to others, and to make progress. He continues to be an inspiration to many.

Paul Nurse

Fred Sanger was an extraordinary scientist as is evident from this informative biographical essay, celebrating his life and his contributions to science. To be awarded one Nobel Prize is very rare, to be awarded two, as was the case for Fred, has only been accomplished by four individuals in over one hundred years. His was a truly outstanding life.

But despite what he achieved Fred was a modest man. He declined a knighthood because he did not wish to be addressed as 'Sir' or to be thought different from others. If you met him late in his life wandering the corridors of the Sanger Institute, you would not think he was one of the greatest scientists of all time, you were more likely to think he was the gardener who had accidentally wandered into the Institute. He was a scientist's scientist, who liked doing experiments and 'messing about in the lab'. Driven by curiosity about the world, he argued that real scientific understanding only came with the highest quality evidence, and so he focused on developing ways to carry out reliable experiments to generate the evidence needed to understand the basic processes underpinning the nature of life. He did not publish much, but what he published was momentous. I doubt if he ever thought about the impact of his work beyond that it should tackle an important question.

So what did Fred achieve? He worked out how to sequence proteins and then how to sequence DNA, both technical tours de force. These experimental procedures had two immensely significant outcomes. The first was that they provided evidence supporting the crucial role that information flow plays in biology. The sequence of nucleotides making up DNA acts as a digital information storage device, and is the basis of heredity. In turn this sequence determines the sequence of amino acids making up proteins, determining their properties to act as catalytic machines carrying out the chemistry in cells which forms the basis of life's processes. The second was to carry out the sequencing of a small bacterial virus, the first fully sequenced DNA-based genome. This proof of principle experiment demonstrated that it would eventually be possible to sequence the human genome, as indeed it was. DNA sequencing is now commonplace and is a major driver for understanding how living organisms work and what goes wrong in disease. For example, look at how it illuminates our own evolution through the sequencing of our ancient ancestors and our understanding of diseases such as cancer.

Fred is one of those few individuals of whom it can be said that they changed the world for ever. This essay gives a personal insight into his science and what he achieved. It allows all of us to be a little closer to this gentle, modest man of genius.

Epilogue

Fred Sanger retired in 1983 aged 65. Many people thought this was too early and that the Director then, Sydney Brenner, should perhaps have persuaded him to stay on at the laboratory. After all he was the most high-profile scientist at the lab. But Sanger himself felt it was the right time to retire. He had reached the pinnacle of his career. He had just won his second Nobel Prize. A third Nobel Prize was out of reach even for a man of Sanger's determination, focus and energy. His work had paved the way for sequencing the human genome. Fred said his capacity for working at the bench and doing experiments was waning and space should be given to someone younger. Fred was to apply his remaining energy to his much larger garden when he and his wife, Joan, moved from Hills Road, Cambridge to 'Far Leys' (named in memory of his childhood home near Birmingham) in Swaffham Bulbeck outside Cambridge in the Fens. Here he was to grow roses, fruit trees, soft fruit, a vine cutting I gave him, along with his many other specimen flowers in his herbaceous borders. He also had time now to spend more time enjoying his hobby of sailing and watching his grandchildren, the children of his second son, Peter, grow up.

The overriding question, the central question in this biography, is why was Fred Sanger so successful as a scientist. What attributes allowed him to succeed twice on two different but fundamental problems and gain two Nobel Prizes? Was it his personal attributes? Was it the influence of his parents and education at school? Was it his own early career in the Department of Biochemistry in Cambridge where he started science? Was it his choice of the scientific problem or his choice of collaborators? Was it in his DNA? Was he perhaps just lucky to be awarded two Nobel Prizes?

There is no doubt from his early childhood he developed an interest in the natural world around him. This interest was cultivated first by his elder brother, Theo, who was fascinated by natural history. Theo was more outgoing, more extrovert than Fred. Fred, only one year younger than Theo, followed his older brother around in their large gardens, first in Rendcomb in Gloucestershire and later in Tanworth near Birmingham. Fred's father, Frederick, and mother, Cicely, must have encouraged this exploratory behaviour in their two sons. But the parents knew their boys were different. Fred was much quieter than the confident, extrovert Theo. This quiet trait in Fred was apparent very early on and was noted by Cicely in her diaries when Fred was less than one year old. Later on Fred found schoolwork easier than Theo, both under the watchful eye of the governess who came to their Tanworth home, and then after that at prep school in Malvern, where Fred was sent as a boarder when he was only nine years old. Fred's mother was generally less concerned with Fred's schooling than with Theo's, as it was Theo who found schoolwork harder. Fred was good at school and caused no concern, although Cicely records that Fred did not manage to get a scholarship to his secondary school at Bryanston.

Did boarding school, which Fred initially disliked, help build Fred's character? His mother thought so. In her diaries she said that Fred became less shy once he went to school. Did it make him more self-sufficient than if he had been sent to a state-supported school near where they lived near Birmingham rather than boarding in the West of England? We shall never know. But we do know that Fred sent his own two sons, Robin and Peter, to boarding school at Bryanston, following in their father's footsteps, suggesting that Fred found the experience of boarding away from home beneficial.

Fred enjoyed a happy, contented and very supportive upbringing. This was the influence of his parents. They were well-to-do, there is no doubt of that, as his father had married the youngest daughter of a wealthy cotton manufacturer. But Fred's parents were also kind and considerate. There were no arguments, just discussions when there were differences of opinions.

Fred must have been a very well-adjusted and an independent boy at school. He had made the conscious decision while still at school that he would not follow his father's advice and study medical subjects at Cambridge. Instead he chose science. But this caused him problems and he had to spend four years as an undergraduate at Cambridge instead of the normal three years to get his BA, because he had made the wrong decision about which subjects to study for the Tripos. Overcoming this obstacle and the death of both of his parents as an undergraduate would have set back many students, but Fred took this in his stride. Perhaps this was in his nature or perhaps it was because the war had started. Everybody had to cope, with whatever happened. It was a time of huge uncertainty. In the early years of the war, Britain was on its own standing up to the Nazis. America was still a reluctant ally, and the formal Allies, France, the Netherlands and Belgium, were quickly overwhelmed by the German invasion all over Europe.

Fred also had to cope with his religious feelings. Brought up as a strict Quaker and staunchly pacifist, he successfully registered as a conscientious objector. Such objectors were unpopular amongst the general population who expected their young men to help in the war effort. His engagement and marriage to Joan Howe in 1939–40 must have helped Fred enormously to survive the uncertainties of the time and his own isolation in Cambridge. His decision to study for a PhD, the degree that set Fred off in his research career, must be the one really critical decision that influenced his life as a scientist. Because he gained a first class degree in biochemistry in his fourth year and clearly had academic potential, the Department of Biochemistry were able to consider Fred for their PhD programme, even though he only applied for this programme unexpectedly, and had to arrange a PhD supervisor at the very last minute.

Fred was now introduced into research and never looked back. His career seemed to blossom from the outset. It was a happy choice for which he was ideally suited. His practical skills acquired from his parents, his brother and his school stood him in good stead. The nature of the problems Fred undertook to solve – how the sequences of the building blocks of proteins and nucleic acids were arranged – was an ideal problem as it challenged his academic and practical skills at solving problems and overcoming them. This challenge suited his temperament. He did not go the route of some scientists in setting up alternative hypotheses and thinking of clever ways of proving or disproving such hypotheses. Many scientists waste time by using this approach. But we must not overlook Fred's personal qualities – attributes he had learned from his father and mother and from his upbringing and schooling. These are primarily his modesty and unwillingness to criticise his colleagues. Rather he always encouraged scientists around him, whether Visiting Professors, his immediate colleagues or his own younger staff such as postdocs, PhD students and laboratory assistants. Moreover he showed, by example, that he was prepared personally to undertake research throughout his entire career. This personal commitment to research inspired everyone who worked with Fred. It was unusual then, and is almost unheard of in science today. That way he maintained loyalty from his staff. This undoubtedly helped him in the last, quite frustratingly slow, stage in solving the structure of insulin, and in the considerable effort his team had to make in completing the DNA sequences of bacteriophages φX174 and λ.

What attributes lead Fred Sanger to achieve the ultimate accolade of obtaining two Nobel Prizes? His personal effort throughout his career was extraordinary. He was focused, committed, undeterred if things did not immediately succeed and determined to succeed in his goals. In my opinion Fred Sanger's two Nobel Prizes speak for themselves. They honour a man who fully deserves both prizes. They place him amongst the few scientists that have achieved this honour – the best known, besides Fred, being Madame Curie. They mark Fred Sanger as one of the great scientists of all time, whose name must be linked to Madame Curie and to Charles Darwin because of the impact of the discoveries they made and will continue to be made in the future.

Appendix: Complete bibliography of Fred Sanger

1. Sanger F. Determination of nucleotide sequences in DNA. *Biosci. Rep.* **24**: 237–253 (2004).
2. Sanger F. The early days of DNA sequences. *Nat. Med.* **3**: 267–268 (2001).
3. Sanger E, Dowding M, eds., *Selected Papers of Frederick Sanger.* Singapore: World Scientific (1996).
4. Sanger F, Nicklen S, Coulson AR. DNA sequencing with chain-terminating inhibitors. 1977. *Biotechnology* **24**: 104–108 (1992).
5. Sanger F. Sequences, sequences, and sequences. *Ann. Rev. Biochem.* **57**: 1–28 (1988).
6. Daniels DL, Sanger F, Coulson AR. Features of bacteriophage λ: analysis of the complete nucleotide sequence. *Cold Spring Harb. Symp. Quant. Biol.* **47**: 1009–1024 (1983).
7. Sanger F, Coulson AR, Hong GF, Hill DF, Petersen GB. Nucleotide sequence of bacteriophage λ DNA. *J. Mol. Biol.* **162**: 729–773 (1982).
8. Anderson S, de Bruijn MH, Coulson AR, Eperon IC, Sanger F, Young IG. Complete sequence of bovine mitochondrial DNA: conserved features of the mammalian mitochondrial genome. *J. Mol. Biol.* **156**: 683–717 (1982).
9. Anderson S, Bankier AT, Barrell BG, de Bruijn MHL, Coulson AR, Drouin J, Eperon, IC, Nierlich DP, Roe BA, Sanger F, Schreier PH, Smith AJH, Staden R, Young IG. In Slonimski PP, Borst P, Attardi G, eds., *Mitochondrial Genes.* Cold Spring Harbor, NY: Cold Spring Harbor Laboratory Press, pp. 5–43 (1982).
10. Sanger F. Determination of nucleotide sequences in DNA. *Science* **214**: 1205–1210 (1981).
11. Anderson S, Bankier AT, Barrell BG, de Bruijn MH, Coulson AR, Drouin J, Eperon IC, Nierlich DP, Roe BA, Sanger F, Schreier PH, Smith AJ, Staden R, Young IG. Sequence and organization of the human mitochondrial genome. *Nature* **290**: 457–465 (1981).
12. Sanger F. Determination of nucleotide sequences in DNA. *Biosci. Rep.* **1**: 3–18 (1981).

13. Sanger F. Nobel Lecture, 1980: Determination of nucleotide sequences in DNA. In *Nobel Prizes, Chemistry 1970–1980*. Singapore: World Scientific, pp. 431–447 (1993).
14. Sanger F, Coulson AR, Barrell BG, Smith AJ, Roe BA. Cloning in single-stranded bacteriophage as an aid to rapid DNA sequencing. *J. Mol. Biol.* **143**: 161–178 (1980).
15. Barrell BG, Anderson S, Bankier AT, de Bruijn MH, Chen E, Coulson AR, Drouin J, Eperon IC, Nierlich DP, Roe BA, Sanger F, Schreier PH, Smith AJ, Staden R, Young IG. Different pattern of codon recognition by mammalian mitochondrial tRNAs. *Proc. Natl Acad. Sci. USA* **77**: 3164–3166 (1980).
16. Air GM, Coulson AR, Fiddes JC, Friedmann T, Hutchison CA III, Sanger F, Slocombe PM, Smith AJ. Nucleotide sequence of the F protein coding region of bacteriophage φX174 and the amino acid sequence of its product. *J. Mol. Biol.* **125**: 247–254 (1978).
17. Sanger F, Coulson AR, Friedmann T, Air GM, Barrell BG, Brown NL, Fiddes JC, Hutchison CA III, Slocombe PM, Smith M. The nucleotide sequence of bacteriophage φX174. *J. Mol. Biol.* **125**: 225–246 (1978).
18. Sanger F, Coulson AR. The use of thin acrylamide gels for DNA sequencing. *FEBS Lett.* **87**: 107–110 (1978).
19. Sanger F, Nicklen S, Coulson AR. DNA sequencing with chain-terminating inhibitors. *Proc. Natl Acad. Sci. USA* **74**: 5463–5467 (1977).
20. Sanger F, Air GM, Barrell BG, Brown NL, Coulson AR, Fiddes CA, Hutchison CA III, Slocombe PM, Smith M. Nucleotide sequence of bacteriophage φX174 DNA. *Nature* **265**: 687–695 (1977).
21. Smith M, Brown NL, Air GM, Barrell BG, Coulson AR, Hutchison CA III, Sanger F. DNA sequence at the C termini of the overlapping genes A and B in bacteriophage φX174. *Nature* **265**: 702–705 (1977).
22. Air GM, Sanger F, Coulson AR. Nucleotide and amino acid sequences of gene G of φX174. *J. Mol. Biol.* **108**: 519–533 (1976).
23. Air GM, Blackburn EH, Coulson AR, Galibert F, Sanger F, Sedat JW, Ziff EB. Gene F of bacteriophage φX174: correlation of nucleotide sequences from the DNA and amino acid sequences from the gene product. *J. Mol. Biol.* **107**: 445–458 (1976).

24. Sanger F. The Croonian Lecture 1975: Nucleotide sequences in DNA. *Proc. R. Soc. Lond. B* **191**: 317–333 (1975).
25. Air GM, Blackburn EH, Sanger F, Coulson AR. The nucleotide and amino acid sequence of the N (5′) terminal region of gene G of bacteriophage φX174. *J. Mol. Biol.* **96**: 703–719 (1975).
26. Sanger F, Coulson AR. A rapid method for determining sequences in DNA by primed synthesis with DNA polymerase. *J. Mol. Biol.* **94**: 441–448 (1975).
27. Sanger F, Donelson JE, Coulson AR, Kössel H, Fischer D. Determination of a nucleotide sequence in bacteriophage f1 DNA by primed synthesis with DNA polymerase. *J. Mol. Biol.* **90**: 315–333 (1974).
28. Sanger F, Donelson JE, Coulson AR, Kössel H, Fischer D. Use of DNA polymerase I primed by a synthetic oligonucleotide to determine a nucleotide sequence in phage f1 DNA. *Proc. Natl Acad. Sci. USA* **70**: 1209–1213 (1973).
29. Jeppesen PGN, Barrell BG, Sanger F, Coulson AR. Nucleotide sequences of two fragments from the coat-protein cistron of bacteriophage R17 ribonucleic acid. *Biochem. J.* **128**: 993–1006 (1972).
30. Sanger F. The eighth Hopkins Memorial Lecture: Nucleotide sequences in bacteriophage ribonucleic acid. *Biochem. J.* **124**: 833–843 (1971).
31. Sanger F, Brownlee GG. Methods for determining sequences in RNA. *Biochem. Soc. Symp.* **30**: 183–197 (1970).
32. Jeppesen PJN, Nichols JL, Sanger F, Barrell BJ. Nucleotide sequences from bacteriophage R17 RNA. *Cold Spring Harb. Symp. Quant. Biol.* **35**: 13–19 (1970).
33. Brownlee GG, Sanger F. Chromatography of ^{32}P-labelled oligonucleotides on thin layers of DEAE-cellulose. *Eur. J. Biochem.* **11**: 395–399 (1969).
34. Adams JM, Jeppesen PG, Sanger F, Barrell BG. Nucleotide sequence from the coat protein cistron of R17 bacteriophage RNA. *Nature* **223**: 1009 1014 (1969).
35. Labrie F, Sanger F. ^{32}P-labelling of haemoglobin messenger and other reticulocyte ribonucleic acids with polynucleotide phosphokinase in vitro. *Biochem. J.* **114**: 29P (1969).
36. Székely M, Sanger F. Use of polynucleotide kinase in fingerprinting non-radioactive nucleic acids. *J. Mol. Biol.* **43**: 607–617 (1969).

37. Adams JM, Jeppesen PG, Sanger F, Barrell BG. Nucleotide sequences from fragments of R17 bacterophage RNA. *Cold Spring Harb. Symp. Quant. Biol.* **34**: 611–620 (1969).

38. Barrell BG, Sanger F. The sequence of phenylalanine tRNA from *E. coli. FEBS Lett.* **3**: 275–278 (1969).

39. Fellner P, Sanger F. Sequence analysis of specific areas of the 16S and 23S ribosomal RNAs. *Nature* **219**: 236–238 (1968).

40. Brownlee GG, Sanger F, Barrell BG. The sequence of 5S ribosomal ribonucleic acid. *J. Mol. Biol.* **34**: 379–412 (1968).

41. Brownlee GG, Sanger F, Barrell BG. Nucleotide sequence of 5S ribosomal RNA from *Escherichia coli. Nature* **215**: 735–736 (1967).

42. Brownlee GG, Sanger F. Nucleotide sequences from the low molecular weight ribosomal RNA of *Escherichia coli. J. Mol. Biol.* **23**: 337–353 (1967).

43. Sanger F, Brownlee GG. A two-dimensional fractionation method for radioactive nucleotides. In Grossman L, Moldave K, eds., *Methods in Enzymology*, vol. XII, Part A, New York: Academic Press, pp. 361–363 (1967).

44. Sanger F, Brownlee GG, Barrell BG. A two-dimensional fractionation procedure for radioactive nucleotides. *J. Mol. Biol.* **13**: 373–398 (1965).

45. Larner J, Sanger F. The amino acid sequence of the phosphorylation site of muscle uridine diphosphoglucose alpha-1,4-glucan α-4-glucosyl transferase. *J. Mol. Biol.* **11**: 491–500 (1965).

46. Marcker K, Sanger F. *N*-formyl-methionyl-s-RNA. *J. Mol. Biol.* **8**: 835–840 (1964).

47. Sanger F, Bretscher MS, Hocquard EJ. A study of the products from a polynucleotide-directed cell-free protein synthesizing system. *J. Mol. Biol.* **8**: 38–45 (1964).

48. Glazer AN, Sanger F. The iodination of chymotrypsinogen. *Biochem. J.* **90**: 92–98 (1964).

49. Glazer AN, Sanger F. Effect of fatty acid on the iodination of bovine serum albumin. *J. Mol. Biol.* **7**: 452–453 (1963).

50. Sanger F, Thompson EO. Halogenation of tyrosine during acid hydrolysis. *Biochim. Biophys. Acta* **71**: 468–471 (1963).

51. Sanger F, Hocquard E. Formation of dephospho-ovalbumin as an intermediate in the biosynthesis of ovalbumin. *Biochim. Biophys. Acta* **62**: 606–607 (1962).

52. Milstein C, Sanger F. An amino acid sequence in the active centre of phosphoglucomutase. *Biochem. J.* **79**: 456–469 (1961).
53. Naughton MA, Sanger F. Purification and specificity of pancreatic elastase. *Biochem. J.* **78**: 156–163 (1961).
54. Naughton MA, Sanger F, Hartley BS, Shaw DC. The amino acid sequence around the reactive serine residue of some proteolytic enzymes. *Biochem. J.* **77**: 149–163 (1960).
55. Sanger F, Shaw DC. Amino-acid sequence about the reactive serine of a proteolytic enzyme from *Bacillus subtilis*. *Nature* **187**: 872–873 (1960).
56. Sanger F. Chemistry of insulin. *Br. Med. Bull.* **16**: 183–188 (1960).
57. Milstein C, Sanger F. The amino acid sequence around the serine phosphate in phosphoglucomutase. *Biochim. Biophys. Acta* **42**: 173–174 (1960).
58. Hartley BS, Naughton MA, Sanger F. The amino acid sequence around the reactive serine of elastase. *Biochim. Biophys. Acta* **34**: 243–244 (1959).
59. Sanger F. Nobel Lecture, 1958: The chemistry of insulin. In *Nobel Lectures, Chemistry 1942–1962*. Amsterdam: Elsevier, pp. 134–146 (1964).
60. Sanger F. Chemistry of insulin: determination of the structure of insulin opens the way to greater understanding of life processes. *Science* **129**: 1340–1344 (1959).
61. Williams J, Sanger F. The grouping of serine phosphate residues in phosvitin and casein. *Biochim. Biophys. Acta* **33**: 294–296 (1959).
62. Harris JI, Naughton MA, Sanger F. Species differences in insulin. *Arch. Biochem. Biophys.* **65**: 427–438 (1956).
63. Brown H, Sanger F, Kitai R. The structure of pig and sheep insulins. *Biochem. J.* **60**: 556–565 (1955).
64. Ryle AP, Sanger F, Smith LF, Kitai R. The disulphide bonds of insulin. *Biochem. J.* **60**: 541–556 (1955).
65. Ryle AP, Sanger F. Disulphide interchange reactions. *Biochem. J.* **60**: 535–540 (1955).
66. Sanger F, Thompson EO, Kitai R. The amide groups of insulin. *Biochem. J.* **59**: 509–518 (1955).
67. Sanger F, Smith LF, Kitai R. The disulphide bridges of insulin. *Biochem. J.* **58**: vi–vii (1954).
68. Ryle AP, Sanger F. Disulphide interchange reactions. *Biochem. J.* **58**: v–vi (1954).

69. Sanger F. A disulphide interchange reaction. *Nature* **171**: 1025–1026 (1953).
70. Sanger F, Thompson EO. The amino-acid sequence in the glycyl chain of insulin. 2. The investigation of peptides from enzymic hydrolysates. *Biochem. J.* **53**: 366–374 (1953).
71. Sanger F, Thompson EO. The amino-acid sequence in the glycyl chain of insulin. 1. The identification of lower peptides from partial hydrolysates. *Biochem. J.* **53**: 353–366 (1953).
72. Sanger F, Thompson EO. The amino-acid sequence in the glycyl chain of insulin. *Biochem. J.* **52**: iii (1952).
73. Sanger F. The arrangement of amino acids in proteins. *Adv. Protein Chem.* **7**: 1–67 (1952).
74. Sanger F, Thompson EO. The inversion of a dipeptide sequence during hydrolysis in dilute acid. *Biochim. Biophys. Acta* **9**: 225–226 (1952).
75. Sanger F, Tuppy H. The amino-acid sequence in the phenylalanyl chain of insulin. 2. The investigation of peptides from enzymic hydrolysates. *Biochem. J.* **49**: 481–490 (1951).
76. Sanger F, Tuppy H. The amino-acid sequence in the phenylalanyl chain of insulin. 1. The identification of lower peptides from partial hydrolysates. *Biochem. J.* **49**: 463–481 (1951).
77. Bailey K, Sanger F. The chemistry of amino acids and proteins. *Ann. Rev. Biochem.* **20**: 103–130 (1951).
78. Sanger F. Some chemical investigations on the structure of insulin. *Cold Spring Harb. Symp. Quant. Biol.* **14**: 153–160 (1950).
79. Sanger F. The chemistry of insulin. *Annu. Rep. Prog. Chem.* **45**: 283–292 (1949).
80. Sanger F. Application of partition chromatography to the study of protein structure. *Biochem. Soc. Symp.* **3**: 21 (1949).
81. Sanger F. Species differences in insulins. *Nature* **164**: 529 (1949).
82. Sanger F. The terminal peptides of insulin. *Biochem. J.* **45**: 563–574 (1949).
83. Sanger F. Fractionation of oxidized insulin. *Biochem. J.* **44**: 126–128 (1949).
84. Sanger F. Some peptides from insulin. *Nature* **162**: 49 (1948).
85. Porter RR, Sanger F. The free amino groups of haemoglobins. *Biochem. J.* **42**: 287–294 (1948).
86. Tiselius A, Sanger F. Adsorption analysis of oxidized insulin. *Nature* **160**: 433 (1947).

87. Sanger F. Oxidation of insulin by performic acid. *Nature* **160**: 295 (1947).
88. Sanger F. The free amino group of gramicidin S. *Biochem. J.* **40**: 261–262 (1946).
89. Sanger F. The free amino groups of insulin. *Biochem. J.* **39**: 507–515 (1945).
90. Neuberger A, Sanger F. The availability of ε-acetyl-D-lysine and ε-methyl-DL-lysine for growth. *Biochem. J.* **38**: 125–129 (1944).
91. Neuberger A, Sanger F. The metabolism of lysine. *Biochem. J.* **38**: 119–125 (1944).
92. Neuberger A, Sanger F. The availability of the acetyl derivatives of lysine for growth. *Biochem. J.* **37**: 515–518 (1943).
93. Harris HA, Neuberger A, Sanger F. Lysine deficiency in young rats. *Biochem. J.* **37**: 508–513 (1943).
94. Sanger F. The metabolism of the amino-acid lysine in the animal body. PhD thesis, Cambridge University (1943).
95. Neuberger A, Sanger F. The nitrogen of the potato. *Biochem. J.* **36**: 662 (1942).

Selected papers by colleagues

1. Smith AJ. The use of exonuclease III for preparing single stranded DNA for use as a template in the chain terminator sequencing method. *Nucl. Acids Res.* **6**: 831–848 (1979).
2. Barrell BG, Bankier AT, Drouin J. A different genetic code in human mitochondria. *Nature* **282**: 189–194 (1979).
3. Friedmann T, Brown DM. Base-specific reactions useful for DNA sequencing: methylene blue-sensitized photooxidation of guanine and osmium tetraoxide modification of thymine. *Nucl. Acids Res.* **5**: 615–622 (1978).
4. Barrell BG, Air GM, Hutchison CA III. Overlapping genes in bacteriophage φX174. *Nature* **264**: 34–41 (1976).
5. Blackburn EH. Transcription and sequence analysis of a fragment of bacteriophage φX174 DNA. *J. Mol. Biol.* **107**: 417–431 (1976).
6. Sedat J, Ziff E, Galibert F. Direct determination of DNA nucleotide sequences: structure of large specific fragments of bacteriophage φX174 DNA. *J. Mol. Biol.* **107**: 391–416 (1976).
7. Blackburn EH. Transcription by *Escherichia coli* RNA polymerase of a single-stranded fragment by bacteriophage φX174 DNA 48 residues in length. *J. Mol. Biol.* **93**: 367–374 (1975).

8. Galibert F, Sedat J, Ziff E. Direct determination of DNA nucleotide sequences: structure of a fragment of bacteriophage φX174 DNA. *J. Mol. Biol.* **87**: 377–407 (1974).
9. Ziff EB, Sedat JW, Galibert F. Determination of the nucleotide sequence of a fragment of bacteriophage φX174 DNA. *Nat. New Biol.* **241**: 34–37 (1973).
10. Ling V. Pyrimidine sequences from the DNA of bacteriophages fd, f1, and φX174. *Proc. Natl Acad. Sci. USA* **69**: 742–746 (1972).
11. Ling V. Fractionation and sequences of the large pyrimidine oligonucleotides from bacteriophage fd DNA. *J. Mol. Biol.* **64**: 87–102 (1972).
12. Dahlberg JE. Terminal sequences of bacteriophage RNAs. *Nature* **220**: 548–552 (1968).
13. Dube SK, Marcker KA, Clark BF, Cory S. Nucleotide sequence of *N*-formyl-methionyl-transfer RNA. *Nature* **218**: 232–233 (1968).
14. Murray K, Offord RE. Use of neutron activation in the characterization of small quantities of nucleic acids. *Nature* **211**: 376–378 (1966).

Notes

Notes to Chapter 1

1. James RR (1994). *Henry Wellcome*. London: Hodder & Stoughton.
2. Mary (May) Willford, *née* Sanger, interview with the author.
3. Cicely Sanger, diary 1918–1937, unpublished, courtesy of Mary Willford, *née* Sanger.
4. Bryanston School DVD, unpublished, courtesy of Mary Willford, *née* Sanger.
5. Silverstein A, Silverstein VB (1969). *Frederick Sanger: The Man Who Mapped Out a Chemical of Life*. New York: John Day.
6. Sanger F (1943) *The metabolism of the amino acid lysine in the animal body*. PhD thesis, University of Cambridge.

Notes to Chapter 2

1. Morton, RA (1969). *The Biochemical Society: Its History and Activities, 1911–1969*. London: The Biochemical Society.
2. Needham J (1962). Frederick Gowland Hopkins. *Perspec. Biol. Med.* **6**: 2–46.
3. Ferry G (1998). *Dorothy Hodgkin: A Life*. Granta Books: London.
4. *Ibid.*
5. Fellows FCI, Lewis MHR (1973). *Biochem. J.* **136**: 329–334.
6. Nos. 90–93 and 95 of Sanger's publications (see Bibliography).
7. Letter from Albert Neuberger to Fred Sanger, dated 4 Sept 1987 (Peter Sanger, family archives).
8. Banting and McCleod were awarded the Nobel Prize in 1923 for this discovery. The decision of the Nobel Committee annoyed Banting, who then shared his cash award with Best, the young medical student who had worked with him.
9. Albumin is a major glycoprotein constituent of blood.
10. Abel JJ (1926). *Proc. Natl Acad. Sci. USA.* **12**: 132–136.
11. Wellcome/Beit/Sanger archive (restricted), 1949–50 report by Sanger to Beit Trustees.
12. There are only 16 (not 20) different DNP-amino acids derived by acid hydrolysis of DNB-tagged insulin, because

methionine and tryptophan are absent in insulin. Glutamic acid and glutamine both give rise to DNP-glutamic acid on acid hydrolysis. Similarly both aspartic acid and asparagine give rise to DNP-aspartic acid.
13. Gordon AH, Martin AJP, Synge RLM (1943). *Biochem J.* **37**: 79–86.
14. Bergmann M, Niemann C (1938). *J. Biol. Chem.* **122**: 577–596.
15. Hirs CHW, Moore S, Stein WH (1960). *J. Biol. Chem.* **235**: 633–647.
16. Hartley BS (1964). *Nature* **201**: 1284–1287; Walsh KA, Neurath H (1964). *Proc. Natl Acad. Sci. USA* **52**: 884–889.
17. Adams MJ *et al.* (1969). *Nature* **224**: 491–495; see also note 10 above.
18. See note 3 above.
19. Sanger F (1988). *Ann. Rev. Biochem.* **57**: 1–28.
20. Menting JG *et al.* (2013). *Nature* **493**: 241–245.
21. Richard Ambler, interview with the author.

Notes to Chapter 3

1. Sanger F (1988). *Ann. Rev. Biochem.* **57**: 1–28.
2. Stimulated by discussions with Chris Anfinsen.
3. Sanger/Wellcome Archive 16, notebook 10, May 1958.
4. Milstein CP (1968). *Biochem. J.* **110**: 127–134.
5. McReynolds L *et al.* (1978). *Nature* **273**: 723–728.
6. Wellcome/Beit/Sanger archive (restricted) 1949–50 report by Sanger to Beit Trustees.
7. Naughton MA *et al.* (1960). *Biochem. J.* **77**: 149–163. Proteases are enzymes that catalyse the degradation of proteins, e.g. those proteins in our food. Serine proteases are proteases in which serine is critical in catalysis and is at the 'active centre' of the enzyme.
8. Sanger F (1988) (no. 5 in Bibliography).
9. Miescher F (1871). *Hoppe-Seylers Med. Chem. Unters.* **4**: 441–460.
10. Avery OT, MacLeod CM, McCarty M (1944). *J. Exp. Med.* **79**: 137–158.
11. Watson JD, Crick F (1953). *Nature*, **171**: 737–738; Wilkins MHF, Stokes AR, Wilson HR (1953). *Nature* **171**: 738–740.
12. Ingram VM (1956). *Nature* **178**: 792–794.
13. Richard Ambler, interview by the author.

14. Volkin E, Astrachan L (1956). *Virology* **2**: 149–161; Jacob F, Monod J (1961). *J. Mol. Biol.* **3**: 318–356; Brenner S, Jacob F, Meselson M (1961). *Nature* **190**: 576–581; Gross F *et al.* (1961). *Nature* **190**: 581–585.
15. Warner JR, Knopf PM, Rich A (1963). *Proc. Natl Acad. Sci. USA* **49**: 122–129.
16. Hoagland MB *et al.* (1958). *J. Biol. Chem.* **231**: 241–257.
17. Crick FHC (1958). *Symp. Soc. Exp. Biol.* **12**: 138–163.
18. Burton K, Petersen GB (1960). *Biochem. J.* **75**: 17–27.
19. Murray K, Offord RE (1966). *Nature* **211**: 376–378.
20. Finch J (2008). *A Nobel Fellow on Every Floor*. Cambridge, UK: Icon Books.
21. *Ibid.*
22. *Ibid.*
23. Holley RW *et al.* (1965). *Science* **147**: 1462–1465.
24. Barrell BG, Sanger F (1969). *FEBS Lett.* **3**: 275–278.
25. Brownlee GG, Sanger F, Barrell BG (1967). *Nature* **215**: 735–736.
26. Brownlee GG, Sanger F (1969). *Eur. J. Biochem.* **11**: 395–399.
27. Adams JM *et al.* (1969). *Nature* **223**: 1009–1014.
28. The related RNA phage sequence of MS2 RNA of 3569 nucleotides was completed in 1976 by Fiers W *et al.* (1976). *Nature* **260**: 500–507.
29. Ziff EB, Sedat JW, Galibert F (1973). *Nature New Biol.* **241**: 34–37.
30. John Sedat, interview with the author.
31. Ling V (1972). *J. Mol. Biol.* **64**: 87–102.
32. See note 18 above.
33. Wu R, Taylor E (1971). *J. Mol. Biol.* **57**: 491–511.
34. A primer (a short single-stranded oligonucleotide) was not needed by Wu and Taylor because it was already provided by the 'sticky', or incomplete, ends of phage λ.
35. Blackburn EH (1975). *J. Mol. Biol.* **93**: 367–374.
36. The primer sequence was ACCATCCA, complementary (corresponding) to Try.Met.Val.
37. See note 29 above.
38. Sanger F *et al.* (1973). *Proc. Natl Acad. Sci. USA* **70**: 1209–1213.
39. *Ibid.*
40. Englund PT (1971). *J. Biol. Chem.* **246**: 3269–3276; (1972). *J. Mol. Biol.* **66**: 209–224.

41. Sanger F, Coulson AR (1975). *J. Mol. Biol.* **94**: 441–448.
42. See note 1 above.
43. Sanger F *et al.* (1977). *Nature* **265**: 687–695.
44. Sanger F, Nicklen S, Coulson AR (1977). *Proc. Natl Acad. Sci. USA* **74**: 5463–5467.
45. Sanger F *et al.* (1978). *J. Mol. Biol.* **125**: 225–246.
46. Sanger F *et al.* (1982). *J. Mol. Biol.* **162**: 729–773.
47. Anderson S *et al.* (1981). *Nature* **290**: 457–465.
48. Barrell BG, Air GM, Hutchinson III, CA (1976). *Nature* **264**: 34–41.
49. Beadle GW, Tatum EL (1941). *Proc. Natl Acad. Sci. USA* **27**: 499–506.
50. Barrell BG, Bankier AT, Drouin J (1979). *Nature* **282**: 189–194.
51. Barrell BG *et al.* (1980). *Proc. Natl Acad. Sci. USA* **77**: 3164–3166.
52. See notes 48 and 50 above.
53. *Selected Papers of Frederick Sanger*, Sanger F, Dowding M (eds.) 1996 copyright, World Scientific Press.
54. Alan Coulson, interview with the author.
55. Rensing UF, Coulson A, Schoenmakers JG (1974). *Eur. J. Biochem.* **41**: 431–438; Rensing UF, Coulson A, Thompson EO (1973). *Biochem. J.* **131**: 605–610; Barrell BG, Coulson AR, McClain WH (1973). *FEBS Lett.* **37**: 64–69.
56. See note 54 above.
57. *Ibid.*
58. M. Susan Brownlee, interview with the author.
59. See note 30 above.
60. See note 1 above.
61. George Petersen, interview with the author.
62. See note 54 above.
63. See note 61 above.
64. See note 1 above.
65. Sanger F, Brownlee GG, Barrell BG (1965). *J. Mol. Biol.* **13**: 373–398.
66. Maxam AM, Gilbert W (1977). *Proc. Natl Acad. Sci. USA* **74**: 560–564.
67. See note 1 above.

Notes to Chapter 4

1. Precipitin reaction: a reaction between a compound – usually a protein – and an antibody to cause a precipitation, observed as a precipitate, hence precipitin. The precipitin indicated the presence and quantity of the compound in question.

2. Sir Alec Jeffreys, inventor of 'fingerprinting' of an individual's DNA profile. 'Fingerprinting', as used by Jeffreys, indicated a complicated pattern of bands derived from the DNA of a person that served to uniquely identify that person in the same way that fingerprints are unique to people. The method defined similarities and differences between related DNA samples. Variation between related individuals, e.g. between brothers in a family, is caused by sequence differences in their genome. The technique is now widely used forensically.
3. School Certificate exams (and later 'O' levels) were the first public examinations taken by children at secondary school (in Fred Sanger's case at Bryanston). They examined basic knowledge in many subjects before specialisation in fewer subjects at a more advanced level in the final two years of secondary education. Nowadays it would not be possible to obtain university entrance simply on the basis of excellent results at the first public examinations.
4. Ruth Ann Sanger, 1918–2001. See Hughes-Jones N, Tippett P (2003). *Biogr. Mems. Fell. R. Soc. Lond.* **49**: 461–473.
5. Robert Leslie Howland was Fred's moral tutor and actually a lecturer in Classics. He was by all accounts an interesting man as he was, at the same time, a member of the British National Athletics Team and represented Britain in the shot-put. He was President of St John's from 1963 to 1967.
6. The Cambridge term 'Tripos' is reputed to go back to the seventeenth century when verses were read out by a person sitting on a three-legged stool (or Tripos) at the graduation ceremony. Eventually the term Tripos was used to refer to the courses offered by the university. Unlike many university courses, the Cambridge Tripos system offers flexibility in the subjects taken by undergraduates, particularly in the Natural Sciences. It encourages a broad-based science course in the first two years (Part I) and the specialisation in the third year (Part II) or fourth year, depending on the subjects taken.
7. Fred Sanger got his BA in 1939 on the basis of his Part I of the Natural Sciences Tripos, awarded a second. He was awarded a first class degree in 1940 in biology for his Part II Biochemistry results.
8. The Peace Pledge Union (PPU) is a British pacifist organisation. It is open to everyone who can sign the PPU pledge: 'I renounce war,

and am therefore determined not to support any kind of war. I am also determined to work for the removal of all causes of war.' Its members work for a world without war and promote peaceful and non-violent solutions to conflict. The PPU emerged from an initiative by Dick Sheppard, Canon of St Paul's Cathedral, in 1934.

Notes to Chapter 5

1. Norman Wingate (Bill) Pirie FRS (1907–97) was a biochemist who, with Sir Frederick Bawden (1908–72), discovered that a virus can be crystallised, by isolating tobacco mosaic virus in 1936.
2. Keto acids contain both carboxyate (CO_2H) and keto (C=O) groups. The simplest keto acid is pyruvic acid ($CH_3.CO.CO_2H$).
3. Glycoproteins are proteins that are modified by the addition of complex carbohydrates linked to particular amino acids such as serine and threonine.
4. Porphyrins (from the Greek word for purple) are a group of complex organic aromatic compounds, many naturally occurring. One of the best-known porphyrins is haem, the pigment in haemoglobin present in red blood cells.
5. Joseph Needham is better known to the general public for his books on the history of science in China, published in the series *Science and Civilization in China.* These books have stimulated interest in the west in the very early contributions of Chinese society to science and technology.
6. Free, or α-amino groups, refer to amino (NH_2) groups forming the N-(or amino-)terminus of the peptide backbone chain of a protein. By contrast, ε-amino groups of lysine are reactive amino groups, wherever lysine occurs in the protein.
7. Methanesulphonyl chloride ($CH_3.SO_2.Cl$) reacts with amino groups in amino acids to form methyl sulphonyl derivatives.
8. Gordon AH, Martin AJP, Synge RLM (1943). *Biochem. J.* **37**: 79–86.
9. Abderhalden E, Stix W (1923). *Hoppe-Seyler's Z. Physiol. Chem.* **129**: 143–156.
10. Sanger F (1945). *Biochem. J.* **39**: 507–515.
11. Gutfreund H (1952). *Biochem. J.* **50**: 564–569.
12. Consden R, Gordon AH, Martin AJP (1944). *Biochem. J.* **38**: 224–232.

13. Landsteiner K (1936). *The Specificity of Serological Reactions.* Berlin: Springer.
14. The γ-globulin fraction of serum contains antibodies.
15. Sanger F (1949). *Biochem. J.* **45**: 563–574.
16. Sanger F, Tuppy H (1951). *Biochem. J.* **49**: 463–481.
17. Dixon M, Webb E (1938). *Enzymes.* London: Longman, Green & Co.
18. Pepsin, chymotrypsin and trypsin are three different proteolytic enzymes.
19. Sanger F, Thompson EOP (1953). *Biochem. J.* **53**: 366–374.
20. The amino acids glutamine and glutamic acid could not be resolved by acid hydrolysis because glutamine is unstable to acid hydrolysis, giving rise to glutamic acid.
21. Sanger F, Thompson EOP, Kitai R (1955). *Biochem. J.* **59**: 509–518.
22. Ryle AP, Sanger F (1955). *Biochem. J.* **60**: 535–540.
23. Ryle AP *et al.* (1955). *Biochem. J.* **60**: 541–546.
24. Brown H, Sanger F, Kitai R (1955). *Biochem. J.* **60**: 556–565.
25. Geiger counters are small hand-held monitors used for detecting radiation.
26. Naughton MA *et al.* (1960). *Biochem. J.* **77**: 149–163.
27. Paper electrophoresis is a technique relying on the separation according to mobility on paper when exposed to a voltage gradient.
28. Milstein C, Sanger F (1961). *Biochem. J.* **79**: 456–469.

Notes to Chapter 6

1. Spackman DH, Stein WH, Moore S (1958). *Anal. Chem.* **30**: 1190–1206.
2. An oligonucleotide is distinguished from a mononucleotide – a single nucleotide, and is a minimum of two nucleotides long. A polynucleotide is many nucleotides in length.
3. Sanger F, Brownlee GG, Barrell BG (1965). *J. Mol. Biol.* **13**: 373–398.
4. Holley RW *et al.* (1965). *Science* **147**: 1462–1465.
5. Brownlee GG, Sanger F, Barrell BG (1967). *Nature* **215**: 735–736.
6. An exonuclease is an enzyme that chops off single nucleotides one at a time from the end of the nucleotide chain. If the exonuclease chops from the left-hand end, as the sequence is usually written, it is called a 5′ exonuclease; if from the right-hand end, it is called a 3′ exonuclease.

7. Ling V (1972). *J. Mol. Biol.* **64**: 87–102.
8. Adams JM *et al.* (1969). *Nature* **223**: 1009–1014.
9. Barrell BG, Sanger F (1969). *FEBS Lett.* **3**: 275–278.
10. Marcker K, Sanger F (1964). *J. Mol. Biol.* **8**: 835–840.
11. Wu R, Kaiser AD (1968). *J. Mol. Biol.* **35**: 523–527.
12. Brownlee GG, Sanger F (1969). *Eur. J. Biochem.* **11**: 395–399.
13. Sanger F *et al.* (1974). *J. Mol. Biol.* **90**: 315–333.
14. Englund PT (1971). *J. Biol. Chem.* **246**: 3269–3276; (1972). *J. Mol. Biol.* **66**: 209–224.
15. Sanger F, Coulson AR (1975). *J. Mol. Biol.* **94**: 441–448.
16. Sanger F *et al.* (1977). *Nature* **265**: 687–695.
17. Sanger F, Coulson AR (1978). *FEBS Lett.* **87**: 107–110.
18. Maxam AM, Gilbert W (1977). *Proc. Natl Acad. Sci. USA* **74**: 560–564.
19. Gronenborn B, Messing J (1978). *Nature* **272**: 375–377.
20. Sanger F *et al.* (1978). *J. Mol. Biol.* **125**: 225–246.
21. Jensen H, Evans EA (1935). *J. Biol. Chem.* **108**: 1–9.
22. Consden R *et al.* (1947). *Biochem. J.* **41**: 596–602.
23. Barrell BG, Air GM, Hutchinson CA III (1976). *Nature* **264**: 34–41.
24. Barrell BG, Bankier AT, Drouin J (1979). *Nature* **282**: 189–194.
25. Anderson S *et al.* (1981). *Nature* **290**: 457–465.
26. Barrell BG *et al.* (1980). *Proc. Natl Acad. Sci. USA* **77**: 3164–3166.

Notes to Chapter 7

1. Smith LM *et al.* (1986). *Nature* **321**: 674–679.
2. Gocayne J *et al.* (1987). *Proc. Natl Acad. Sci. USA* **84**: 8296–8300.
3. Recollection by the author, who was present at the opening ceremony.
4. Sulston J, Ferry, G (2002). *The Common Thread.* London: Bantam Books.
5. The *C. elegans* Sequencing Consortium (1998). *Science* **282**: 2012–2018.
6. Lander ES *et al.* (2001). *Nature* **409**: 860–921; Venter JC *et al.* (2001). *Science* **291**: 1304–1351.
7. Adams J, interview with the author, 2011.
8. The ENCODE Project Consortium (2012). *Nature* **489**: 57–74.

9. Promoters are relatively short (<20 bases) DNA sequences that specify the beginning of genes, i.e. where the RNA polymerase binds to start transcription. Enhancers are short (<20 bases) DNA sequences – often distant from promoters – which, like promoters, bind transcription factors and enhance the efficiency of promoters. DNase I hypersensitivity sites are regions of the genome which are accessible to DNA cleavage by the enzyme, DNase I. They mark open, chromatin regions, where nucleosomes are absent, and correlate with transcription start sites.
10. Bentley DR *et al.* (2008). *Nature* **456**: 53–59.
11. The 1000 Genomes Project Consortium (2012). *Nature* **491**: 56–71.
12. Saunders CJ *et al.* (2012). *Sci. Transl. Med.* **4**: 154ra135.
13. Wikipedia entry for *polymerase chain reaction.*
14. Dressman D *et al.* (2003). *Proc. Natl Acad. Sci. USA* **13**: 8817–8822.
15. Fedurco M *et al.* (2006). *Nucl. Acids Res.* **34**: e22.
16. Ronaghi M *et al.* (1996). *Anal. Biochem.* **242**: 84–89.
17. See note 10 above.
18. Rorhberg JM *et al.* (2011). *Nature* **475**: 348–352.
19. Eid J *et al.* (2009). *Science* **323**: 133–138.
20. Shendure J *et al.* (2005). *Science* **309**: 1728–1731.
21. Stoddart D *et al.* (2009). *Proc. Natl Acad. Sci. USA* **106**: 7702–7707.
22. Pseudogenes are near-repeat, non-functional versions of genes, first reported by the author in Jacq C, Miller JR, Brownlee GG (1977). *Cell* **12**: 109–120.
23. Hutchinson III CA (2007). *Nucl. Acids Res.* **35**: 6227–6237; Shendure JA, Porreca GJ (2011). *Curr. Protocols Mol. Biol.* 7.1.1–7.1.23; Shendure J, Hanlee J (2008). *Nature Biotechnol.* **26**: 1135–1145; Metzker ML (2010). *Nature Genet.* **11**: 31–46.

Notes to Chapter 8

1. Fred Sanger, in an interview by Newell J (1992). *Helix: Amgen's Magazine of Biotechnol.* **1**: 4–9.
2. Fred Sanger, interview by the author, 1992 (see above, p. 128).
3. See note 1 above.
4. Payling Wright G (1956). *Introduction to Pathology*, 2nd edn. London: Longman, Green & Co.

5. Broca P (1866). *Traité des tumeurs.* Paris: Asselin.
6. Lane-Claypon JE (1926). *A Further Report on Cancer of the Breast*, Report on Public Health and Medical Subjects no. 32. London: HMSO.
7. Heilman FR, Bittner JJ (1944). *Cancer Res.* **4**: 578.
8. Miki Y *et al.* (1994). *Science* **266**: 66–70.
9. Wooster R *et al.* (1995). *Nature* **378**: 789–792.
10. *TP53* codes for a transcription factor (a protein that can bind to DNA), that regulates the expression of many other genes. The *TP53* gene is mutated in many cancers, in addition to breast cancer, and is believed to have many functions in the cell including the control of cell division. It is a classic tumour suppressor gene, which when mutated allows tumour growth. Most, but not all, of the mutations occur in its DNA-binding domain. Tumour suppressor genes, or anti-oncogenes, are genes whose protein product helps to prevent cancers arising. When the gene is mutated that predisposes to cancer.
11. Stephens PJ *et al.* (2012). *Nature* **486**: 400–404; Shah SP *et al.* (2012). *Nature* **486**: 395–399; The Cancer Genome Atlas Network (2012). *Nature* **490**: 61–70; Curtis C *et al.* (2012). *Nature* **486**: 346–352; Ellis MJ *et al.* (2012). *Nature* **486**: 353–360; Banerji S *et al.* (2012). *Nature* **486**: 405–409.
12. The seven most commonly mutated genes in breast cancer, according to Stephens *et al.* (see note 11 above), are *TP53*, *PIK3CA*, *ERBB2* – coding for the HER2 protein, *MYC*, *FGFR1/ZNF703*, *GATA3* and *CCND1*. *TP53*, *MYC* and *GATA3* are genes coding for proteins that bind directly to DNA. cMyc, the protein encoded by the *MYC* gene, is a transcription factor for many genes. Its importance in cellular regulation is illustrated by the fact that it is one of only four transcription factors required to reprogramme differentiated cells to become pluripotent, i.e. they have the ability to differentiate into any cell type. cMyc is one of the so-called Yamanaka factors (Takahashi K, Yamanaka S (2006). *Cell* **126**: 663–676). Gata-3, coded by the *GATA3* gene, is another very important transcription factor that is needed for the development of mature mammary cells from mammary stem cells (Kouros-Mehr H *et al.* (2006). *Cell* **12**: 1041–1055). *ERBB2* and *PIK3CA*

are involved in receptor binding and cell signalling, *FGFR1/ZNF703* in signalling and regulation of transcription, and *CCND1* in the cell cycle.

13. Kandoth C *et al.* (2013). *Nature* **502**: 333–339.
14. Davies H *et al.* (2002). *Nature* **417**: 949–954.
15. Kinases are enzymes that add phosphate, i.e. phosphorylate, substrates, usually through ATP. B-Raf belongs to the serine/threonine group of protein kinases. A kinase domain is a region of a protein that is involved in kinase activity or in activating kinase function.
16. Tsai J *et al.* (2008). *Proc. Natl Acad. Sci. USA* **105**: 3041–3046.
17. Chapman PB *et al.* (2011). *N. Engl. J. Med.* **364**: 2507–2516.

Notes to Chapter 9

1. Sanger F (1988). *Ann. Rev. Biochem.* **57**: 1–28.

Index

Numbers in bold refer to figures.

Printed in the United States
By Bookmasters